SCHOTT GUIDE TO GLASS

SCHOTT GUIDE TO GLASS

Heinz G. Pfaender

Revised and Expanded
by Hubert Schroeder

VAN NOSTRAND REINHOLD COMPANY
NEW YORK   CINCINNATI   TORONTO   LONDON   MELBOURNE

Originally published under the title *SCHOTT GLASLEXIKON*
by mvg Moderne Verlags GmbH. German edition
copyright © 1980 by Heinz G. Pfaender and Hubert Schroeder.

Copyright © 1983 by Van Nostrand Reinhold Company Inc.

Library of Congress Catalog Card Number: 82-23899
ISBN: 0-442-27432-7
ISBN: 0-442-27435-1 pbk.

Manufactured in the United States of America

Published by Van Nostrand Reinhold Company Inc.
135 West 50th Street, New York, N.Y. 10020

Van Nostrand Reinhold Publishing
1410 Birchmount Road
Scarborough, Ontario MIP 2E7, Canada

Van Nostrand Reinhold
480 Latrobe Street
Melbourne, Victoria 3000, Australia

Van Nostrand Reinhold Company Limited
Molly Millars Lane
Wokingham, Berkshire, England

15 14 13 12 11 10 9 8 7 6 5 4 3 2 1

Library of Congress Cataloging in Publication Data

Pfänder, Heinz G.
  Schott guide to glass.

  Translation of: Schott Glaslexikon.
  Includes index.
  1. Glass. I. Schroeder, Hubert. II Title.
TP857.P4513 1983   666'.1   82-23899
ISBN 0-442-27432-7
ISBN 0-442-27435-1 (pbk.)

Foreword

The manifold forms and uses of glass are becoming increasingly important in science, industry, and our personal lives. This constantly improving material interests a range of people extending beyond the relatively small number of glass experts. Naturally, questions arise as a result of this widespread interest. For this reason, we have heeded the publisher's suggestion to develop a glass primer which answers many questions and explains much of the terminology.

The bases for this Schott Guide to Glass were the lecture manuscript, "Glass Science for Designers" by Prof. Dr. -Ing. Heinz Pfaender of the Darmstadt Technical College, and the Schott pamphlet, *Concepts of Technical Glass from A to Z*. With the aid of his co-workers, Prof. Dr. rer. nat. Hubert Schroeder, who headed our Central Department of Research and Development until his retirement, wrote the manuscript which evolved into this book. The Osram Company was kind enough to provide the section entitled "Glasses for Lamps." We thank all participants for helping to bring about this reference work.

The *Schott Guide to Glass* will give experts, interested amateurs, and those who work with glass a glimpse into the diversity of this fascinating material.

Mainz, West Germany
Schott Glaswerke

Introduction

Glass is possibly the oldest man-made material, used by man without interruption since the beginning of recorded history. Unlike bronze or iron, however, it has not lent its name to any historical epoch. Still, the use of glass from handblown goblets to electronic components has grown with the rise of the industrial era and greatly affects present life. Glassmaking has always been one of the few truly integrated manufacturing processes where native minerals are transformed into an incredible variety of finished products within a single factory. Although glass production is an energy intensive process with fuel requirements which have changed the ecology and economy of large areas in past times, it uses common and virtually inexhaustible raw materials—in sharp contrast to the relatively scarce metallic ores.

The uses of glass are manifold and modern technologists are almost continuously proposing new applications. Glass is replacing less abundant and more expensive materials as witnessed by the growing importance of glass fibers for communication purposes. Nuclear technology would be unthinkable without radiation shielding apparatus made of glass, and there are many other areas where recently developed glasses are irreplaceable.

Perhaps the preceding millenia have not had a Glass Age because it is still to come. Scientists are already discussing metals and other materials which can be transformed into a glassy state, thereby opening up many new possibilities. Glass—a puzzling material? The *Schott Guide to Glass* contains the answers to most of the questions.

Picture Credits

The authors would like to express their thanks to *Schott Glaswerke*, Mainz, West Germany and to the following German firms for providing pictures:

Bavaria-Boote Hans Bösch, Bad Endorf
Flachglas A.G., Gelsenkirchen
Gerresheimer Glas AG, Düsseldorf
Glaswerk Schuller GmbH, Wertheim
Informations-Zentrum Glas GmbH, Düsseldorf
Osram GmbH, Berlin and Munich
Rheinisches Landesmuseum, Trier
Sekurit-Glas Union GmbH, Aachen
Verlag Chemie GmbH, Weinheim
Wallstab, Kurt, Darmstadt
Wella AG, Darmstadt

Contents

1. The History of Glass

Nobody knows exactly when glass was first made. The oldest finds date to 7000 B.C. in the later Neolithic period.

Glass originated in the Near East. The earliest finds are in Egypt and in Eastern Mesopotamia (Iraq). Glassmaking also developed independently in Mycenae (Greece), China, and North Tyrol.

Ancient glass manufacture is believed to be closely related to pottery making, which was flourishing in Upper Egypt about 8000 B.C. While firing pottery, the chance presence of calciferous sand combined with soda and the overheating of the pottery kiln may have resulted in a colored glaze on the ceramics.

Not until 1500 B.C. was glass produced independently of ceramics and fashioned into separate items.

Other theories consider glass to be a by-product of bronze smelting. Actually, glass and bronze often appear jointly in cultural history since there are close technical ties between the melting process of these two oldest artistic materials.

GLASS IN EGYPT

For a long time, the development of glass was controlled by the status of melting techniques. Very few people mastered the art of glassmaking. The furnaces used in bronze smelting and pottery making were too simple to melt a bubble-free, easily shaped glass.

Around 3000 B.C., Egyptian glassmakers systematically began making pieces of jewelry and small vessels. They modeled viscous glass, cooled to 900°C around a solid mold made of sand or clay.

About 1500 B.C., the Egyptian glassmakers developed their technique an important step further. In order to make vessels for salves or oils, a ceramic negative core was used. A craftsman dipped the form attached to a rod into molten glass, and the first usable hollowware was created.

Constant rotation of the core in the molten glass caused the glass to adhere to the form. Rolling it on a flat stone slab smoothed the surface. Incised or shaped slabs allowed the surface to be decorated. Handles or carrying rings were then added. (See Fig. 1-1.)

Available raw materials were used to melt glass. The addition of copper or cobalt compounds to the glass composition yielded blue tints. Glasses having a brownish appearance were also found. The clay tablet library of the Assyrian King Ashurbanipal (669–626 B.C.) contains cuneiform texts with glass formulas, the oldest of which reads approximately: "Take 60 parts sand, 180 parts ashes of sea plants, 5 parts chalk—and you will get glass." This glass composition contains essentially all the materials used today, although the proportions are somewhat inexact. The low amount of sand, however, leads to the conclusion that achievable melting temperatures during the last millenium before Christ were still not very high, and it was only possible to make soft glass suitable for fashioning simple vessels and other wares.

In the course of the centuries, the art of glassmaking continued to spread. There were so many glassmaking operations in the Nile Valley from Alexandria to Luxor that one can speak of a glass industry of sorts. Similar developments occurred between the Tigris and Euphrates in Iraq, in Syria, on Cyprus, and on Rhodes.

By 1000 B.C., the glassmakers in the Eastern Mediterranean area and bordering regions were creating larger vessels and bowls through the development of new processes. For example, glass rods made of

Fig. 1-1. Sand core vessel. Egypt, ca. 1000 B.C.

multicolored fibers were sliced and placed in molds, and glass was poured into the intervening spaces. Simple casting and pressing methods were also known. But the possibilities for the production of flat and deep bowls were exhausted.

A REVOLUTION IN TECHNOLOGY: THE GLASSBLOWING PIPE

Around 200 B.C. Syrian craftsmen in the area between Sidon and Babylon made a decisive technical breakthrough with the discovery of the glassblowing pipe. This tool is made from an iron tube about 100 to 150 cm long with an opening about 1 cm in diameter. It has an insulated handle with a mouthpiece at one end and a button-like extension at the other end. The glassmaker uses it to get a gob of molten glass from the furnace and blows it into a hollow body. Since this time, the glassblower's pipe has only been slightly improved, despite technological progress.

The blowing of glass with a pipe enabled not only simple, round vessels to be made, but also thin-walled, fine, multishaped glasses. (See Fig. 1-2.) Blowing into a wooden mold allowed products to be standardized and duplicated. Depressions fashioned into the molds, such as ribs, diamonds, or nets, created decorations on the surfaces of the glasses. The application of the glassblowing pipe was also the preliminary step for making flat glass. In order to do this, glass was

Fig. 1-2. Roman spherical vase.

blown into large cylindrical bodies, then cut up, and "ironed" flat while still in the hot state.

The well-developed trade relations among the peoples of the Roman Empire, its highway and transport networks, and a Roman administration conducive to economic progress, were ideal prerequisites for the quick spread of the new invention and the art of glassmaking. In all parts of the Empire, from Mesopotamia to the British Isles, from the Iberian Peninsula to the Rhine, glassworks were founded. The craft experienced its first period of blossoming. Pliny the Elder (23–79 A.D.) described the composition and manufacture of glass in his encyclopedia, *Naturalis Historia*.

GLASS IN THE PERIOD OF THE ROMAN EMPIRE

In Alexandria around 100 A.D., the introduction of manganese oxide into the glass composition combined with improved furnaces resulted in the first successful attempt to make colorless glass. The ability to produce higher temperatures and to maintain better control over combustion atmosphere improved the quality of the glass due to more complete melting of the constituent materials.

The ostentation of the Roman emperors provided a further impetus to glassmaking. Artistically decorated luxury glasses with filigree, mosaic, and engraved decors came into fashion. Glass was made into jewelry and used for imitations of precious stones. The antique art of glass coloring flourished. (See Fig. 1-3.)

The Roman glassworks usually arose in the vicinity of suitable deposits of sand. Labor imported from Alexandria played an important role. Until the Middle Ages soda was imported from Egypt and Syria. There were large numbers of glassworks in the Campagna, as well as in Rome proper. Roman glassworks owners began identifying their products with their firms' logos as early as the first century A.D., and they sold them throughout the Empire. Roman glass specialties were shipped as far as China via the silk routes, although glass had already been developed there independently.

Fig. 1-3. Diatrate vase.

FROM LUXURY PRODUCT TO EVERYDAY ITEM

Numerous glassworks were in operation in the Syrian centers of Sidon and Tyre, in Egyptian Alexandria, in East Roman Byzantium, in Aquileia in Northern Italy, in the northern French cities, Amiens and Boulogne, and in the Germanic cities of Cologne and Trier. (See Fig. 1-4.)

Fig. 1-4. Cabbage stem glass.

For a long time, polished sheets of copper or silver served as mirrors. Then the Phoenicians created small glass mirrors with tin underlays. Since the flat glass they used did not have a level surface, glass mirrors posed no great threat to metal mirrors for hundreds of years. It was not until the thirteenth century in Germany, when the back side of a flat piece of glass was coated with a lead-antimony layer, that a quality mirror was successfully made from glass. This discovery was later improved by the Venetian artisans, but the mirror format remained essentially unchanged.

Only after the discovery of the plate pouring process in 1688 in France under Louis XIV, could large-surface mirrors be created. For this purpose, glass was flattened on a pouring table with rollers, and after cooling, the surfaces were ground and polished smoothly and evenly. Thus plate glass, a flat glass of the highest quality, appeared. By coating this glass with a low melting point metal, it was made into mirrors.

An ancient, long unfulfilled desire was to provide a transparent material for house windows. Since antiquity, parchment and oiled linen had to suffice as coverings for small window openings. Glazed windows were considered a great luxury until well into the Middle Ages.

For centuries window panes were blown with a glassblowing pipe, cut up, and rolled flat. The window dimensions were very small because the glassmaker could only handle a limited quantity of glass. The *bull's-eye* pane appeared in France during the fourteenth century. Its name is derived from the hub-like spiral, the bull's-eye, in the center. To make it, the glassblower first blew a glass ball which was then opened opposite from where the glass was attached to the pipe. The glass was bent outward and spun flat. The finished panes had diameters up to 15 cm and were joined together with lead strips and made into windows.

Among the oldest buildings with windows in Germany are the tenth century Tegernsee Cloister and the eleventh century Augsburg Cathedral with its five prophet windows.

The peak period of stained glass began during the fifteenth century. Churches, palaces, town halls, guild halls, inns, and private houses had glass windows decorated with historical scenes or coats of arms. The spread of stained glass was presumably a direct result of the high windows of Gothic cathedral architecture. The use of colored glass moderated interior light levels and provided an appropriately moving ambience.

THE ROLE OF VENICE

In the Middle Ages, the old merchant metropolis of Venice eventually evolved into the center of Western glassmaking art. At one point, more than 8,000 people were supposedly employed in the Venetian glass industry. The merchants of Venice dominated trade in the Mediterranean and between the fifteenth and seventeenth centuries glassmaking in the city reached its peak, not only the actual glassmaking, but the finishing work as well.

Venetian glassware designers were strongly influenced by many aspects of Islamic art. Syrian enamel painting was further developed by them. The apex of Venetian glassmaking artistry was the creation of the purest crystal glass characterized by an inimitable glitter and absolute clarity. Pure quartz sand and potash made from sea plants were necessary for it. (See Fig. 1-5.)

Fig. 1-5. Gold-ruby bottle, Venice, late 17th Century.

Goblets with hollow stems and footed vases with reliefs of the lion's head of St. Mark characterized the peak of Venetian glass artistry. Bizarre winged glasses and cut cased glass mark the decline of the glass making art of the Renaissance in the seventeenth century. The cutting techniques and decorations exhibit a highly developed technology, but their effect is overdone and effete.

Glassmakers in northern Europe, primarily in the Low Countries and Germany, took up the Venetian tradition and spearheaded a transition to moderate design.

Venice jealously guarded its glass recipes, especially that for crystal glass. At one point, the glassmakers, who were housed on the island of Murano, faced the threat of death if they disclosed a formula. These master craftsmen held positions of high prestige, and not uncommonly attained ranks of nobility.

GLASS IN GERMANY

If glassworks which were founded and operated in Germany throughout the Roman period and then disappeared are disregarded, then one can begin to speak of the beginning of a German glass industry during the Middle Ages. Expatriate Venetian glassmakers founded glassworks and worked in them also. They produced glasses in the Venetian style.

German glassmakers settled in remote areas of central Germany. An increasing quantity of glass was being produced in the Spessart Mountains, the Thuringian Forest, the Solling Mountains, the Black Forest, the Bavarian Forest, the Fichtel Mountains, the Bohemian Forest, the Erz Mountains, the Riesen Mountains, and the Iser Mountains. Initially, a greenish sand- and potash-based glass was melted. Potash (potassium carbonate) was obtained from beech and oak wood. The tree trunks were burned, and the ashes were leached in containers, the so-called pots (hence, the name *potash*). The forests also served as sources of fuel for the glass furnaces. The finished product was called forest glass or *Waldglas*. Most German and Bohemian glasses of the Middle Ages were made of *Waldglas*.

When the surrounding forests were cut down, the glassworks (usually only quickly constructed wooden sheds for the furnace and for storage of finished glasses) were relocated. (See Fig. 1-6.)

A good example of medieval German glass production is found in the Bavarian Forest. Its glass history is interesting because glass production remains the dominant industry in this region today, and the glass industry in other areas of Germany evolved similarly.

These glassworks had a wide range of products: drinking glasses, bull's-eye windows, flat glass for the guilds of the mirror makers in Nuremburg and Augsburg, and glass beads for rosaries. Glass beads from the Bavarian Forest were sold as far away as Spain.

Following the tumult of the Thirty Years' War, which resulted in the expulsion of the predominately Protestant glassmakers, the glass industry was quickly reestablished, and the social standing of glassmakers increased. Near the end of the seventeenth century, there were 60 glassworks in the Bavarian Forest.

Conditions worsened again in the first half of the nineteenth century. The forest was no longer able to provide the required amount of potash and firewood. Raw materials had to be transported from afar. In 1877, the introduction of railroad lines into the region catapulted it into the Industrial Age.

Fig. 1-6. Model of an agricola oven.

ON THE PATH TO GLASS TECHNOLOGY

The entire history of glass is characterized by the efforts of individuals who perfected and further developed production processes and products.

In 1679, Johann Kunckel (1630–1703), the director of the glassworks established near Potsdam by Friedrich Wilhelm of Prussia, the Great Elector, included texts from his own experience and those of others in his handbook, *Ars vitraria experimentalis*. This publication was recognized as the scientific basis of the German art of glassmaking until the nineteenth century.

In Munich, Joseph Fraunhofer (1787–1826), the son of a glass master and trained mirror maker, intensively studied glassmaking technology. After countless attempts, he succeeded in producing glasses with qualities suitable for optical instruments. His telescopes and microscopes were renowned. In 1823, he was named Professor of Physics and was later granted noble status.

In 1676, English glassmakers developed lead crystal. The addition of lead oxide to the glass formula yielded a glass of high brilliance and pure ring. It was very suitable for deep cutting. This was not achievable on the continent for another hundred years. High purity lead glass was used as flint glass for optical purposes.

The promotion of lignite and hard coal as fuels and the establishment of a soda industry released the glass industry from dependence on wood. The glassworks no longer had to be located in remote forested regions. The only requirement was an area with available transport facilities.

The pot furnaces (in which ceramic pots containing raw materials were placed) which had been used since antiquity, were not sufficient for mass production. The discovery of the tank furnace, which could hold up to several hundred tons of material, allowed continuous production and the use of machinery. Furnace technology was improved by the regencrative process, in which the exhaust heat from the melting furnace warms the gaseous fuel and fresh air prior to combustion, so that the fuel is more efficiently utilized and higher melting temperatures are achieved.

Shortly before 1900, the American Michael Owens (1859–1923), invented the automatic bottle blowing machine which was introduced in Europe after the turn of the century. Somewhat later, processes for mechanical production of flat glass were available, without which the quickly rising demand for architectural glass could not have been met. Three hundred thousand standardized glass panes were used as wall panels for the Crystal Palace built by Paxton in London in 1851 for the World Exhibition. This was one of the earliest examples of the use of glass as a structural material.

ERNST ABBE AND OTTO SCHOTT

Two German scientists laid the foundation for modern glass technology. Otto Schott (1851–1935), a chemist and technologist from a family of glassmakers, investigated the dependence of the physical qualities of glass on its composition by using scientific methods. In his father's basement laboratory, he studied the influence of many chemical elements on glass. In a manner of speaking, glass was rediscovered.

In 1876, Otto Schott contacted Ernst Abbe (1840–1905), who was a professor at the University of Jena and co-owner of the Carl Zeiss firm. (See Figures 1-7 and 1-8.) Abbe needed suitable glasses for his high-quality optical instruments. The glasses had to be free from defects and of the highest purity. They also had to have certain qualities,

Fig. 1-7. Otto Schott.

Fig. 1-8. Ernst Abbe.

as required lenses used so far suffered from the so-called secondary spectrum associated with previous glass types. (See Fig. 1-9.)

After years of disappointing attempts, Otto Schott succeeded in producing a glass of the required quality on his ninety-third trial melt.

He moved to Jena, and in 1884, along with Ernst Abbe, Carl Zeiss, and Zeiss' son Roderick, established the *Glastechnisches Laboratorium Schott und Genossen* which later became the *Jenaer*

```
Translation:


                                        Witten, May 27, 1879


Professor Dr. Abbe
Jena


Dear Sir:


  I recently produced a glass, in which a considerable amount of
lithium was introduced, and the specific gravity of which was
relatively low. I suspect that such a glass will exhibit
excellent optical properties, and I therefore wanted to inquire
whether you or one of your colleagues might be willing to test
it for refractive index and dispersion to determine whether my
above supposition is correct. Your connections with Zeiss
should make it easy for you to have the necessary grinding and
polishing of the glass done. If you should be agreeable to my
proposal, I would gladly send you several samples of the glass.


                              With deepest respect
                              and regards,



                              Otto Schott


Address: Dr. Otto Schott
         Witten
         Westphalia
```

Fig. 1-9. The first letter from Otto Schott to Ernst Abbe, dated May 27, 1879.

Glaswerk Schott & Gen. Schott, the 33 year old chemist, then devoted himself solely to glass research.

One of the first new products was a thermometer glass which hardly expanded when heated and therefore did not affect measurement precision. Other new types of glass and melting processes were conceived and investigated: industrial glasses capable of withstanding heat, pressure, and corrosion; optical glasses for microscopes and telescopes in small and large configurations; glasses for high-power camera lenses. Over the years, there were hardly any areas of industry which were not supplied with quality glasses from Jena. Heat-resisting glass for cooking and baking became a common household item.

The Jena Glass Works soon had a worldwide reputation. It soon became evident that its founders also thought of their social responsibilities. "I do not intend to die a millionaire," declared Abbe. He transferred his wealth to the Carl Zeiss Foundation which he had founded. Otto Schott later added his share of the business to the foundation. The workers were assured of secure positions, shared the profits, and along with their families were cared for in cases of illness, disability, and death. The University of Jena received financial support. The firm was one of the few in the world to introduce the 8-hour work day in 1900.

In 1945, shortly after the end of World War II, the American army transferred key personnel from the Zeiss and Schott firms to West Germany. The present day headquarters of the Schott Group, the *Schott Glaswerke*, was set up in Mainz in 1952. The seat of the Zeiss Foundation was moved from Jena, East Germany to Heidenheim, West Germany. The Schott Group is the European leader in the development, manufacture, and sales of special glasses. Schott commands worldwide leadership in several areas, including optical glass. With approximately 12,000 employees in the Federal Republic of Germany, the group of companies manufactures more than 50,000 articles and has subsidiaries and agents in about 100 countries. The main industrial areas served are electrotechnology and electronics; optics and instrumentation; chemistry and pharmacy; information transmission and traffic technology; consumer products; and architecture.

GLASSMAKING IN THE UNITED STATES (Rough Outline)

  I. Mixed Traditions of English and German Glassmaking
 II. Earliest Efforts Were Colonial Enterprises
     1. Jamestown—failed enterprise. 1600s
     2. Annealing—1740s
     3. Wistar—South Jersey Waldglas Industry. 1780s
III. Glassmaking in the New Nation—Three Major Centers.
    19th Century
     1. South Jersey—Waldglas, artistic ware
     2. Pittsburgh, Ohio—Frontier Glassmakers
       Pittsburgh, Ohio—Emphasis on energy and
                    potash—forest products and
                    NC1 utilized.
     3. Northern New York/New England—Glasshouses
       (Boston and Sulwich—Ressel glass)
       Technical innovations for artistic and consumer
       glass—forerunners of Corning.
 IV. Glassmaking in 20th Century
     1. Growth of American optical glass industry—Corn-
       ing, Bausch and Lomb, American Optical, Pitts-
       burgh Plate Glass.
     2. Industrial and Consumer glasses—Owens-Illinois,
       PPG, Libbey, Others.

GLASS THE WORLD OVER

Simple forms of glass are produced worldwide. The most important raw materials and heating fuels are available nearly everywhere, and the necessary technology can be easily transferred. Very few civilized countries do not produce glass. The building of manufacturing facilities for glass containers for food, drinks, and household use usually marks the beginning of industrialization in the developing countries.

Thus, more people are becoming part of a glass tradition which can be traced back for thousands of years. There are no indications that this trend will soon change, for the raw materials needed to make glass are plentiful. In fact, glass has shown the potential to replace many materials which are becoming scarce.

2. Glass, the Material

WHAT IS GLASS?

When considered as a material, glass is a collective term for an un-limited number of materials of different compositions in a glassy state. Glassy materials can also occur naturally. For example, obsidian, often found in volcanic areas, has a composition comparable to man-made glass. It consists of sand, and sodium and calcium compounds; and it was fashioned into knives, arrowheads, spearheads, and other weapons in antiquity. Natural glass in the form of obsidian was chiefly used by the peoples of the Eastern Mediterranean. It was much in demand in this region as an object of trade. The Aztecs in Mexico were also familiar with obsidian and made religious and household items out of it.

Many chemical materials possess the capability for forming glass. Chief among the inorganic materials are the oxides of silicon (Si), boron (B), germanium (Ge), phosphorus (P), and arsenic (As). If they are cooled quickly after melting, they solidify without crystallization. Glass is created.

These glass formers also exhibit this behavior when mixed with other metallic constituents within certain system-dependent compositional limits. The addition of such "glass modifying" components changes bonding relationships and structural groupings resulting in changes in the physical and chemical characteristics of the glasses. The glassy state is not limited to oxides, however, it also appears upon rapid cooling of several sulphur and selenium compounds. Under some extreme circumstances, glass can be made from certain oxide-free metallic alloys. Many organic liquids also transform into the glassy state at low temperatures (such as glycerine at $-90°C$).

Until the eighteenth century, glass was exclusively made from sand, soda, potash, and lime. Coloring metallic oxides were also occasionally added. Today, about 60% of the 90 or so naturally occurring elements from hydrogen to uranium are used in the manufacture of glass. There are even some glass types (optical glasses, for example) which require almost 20 different ingredients. The possibilities for the applications of glass in science and technology are therefore virtually boundless.

Scientists have several answers to the question: "What is glass?" One of the most popular is: "Glass is an inorganic product of melting, which when cooled without crystallization, assumes a solid state." Or: "a frozen supercooled liquid is called glass."

Actually, glass acts as an extremely viscous liquid which deforms very slowly under external force at normal temperatures. The naked eye cannot detect the deformation, but there are scientific procedures to calculate and measure it.

The following definition is more precise: *Glass includes all materials which are structurally similar to a liquid but which have a viscosity so great at normal ambient temperatures that they can be considered as solids. In a more limited sense, the term "glass" denotes all inorganic compounds which possess these basic qualities.* At the same time, a distinction is drawn in comparison to plastics. They are organic in nature and should never be designated as glass even if they are transparent.

GENERAL CHARACTERISTICS OF THE GLASSY STATE

Physically, all glasses are energetically unstable compared to a crystal of the same composition. In general, when cooling a melted substance, crystallization should begin to occur when the temperature falls below the melting point (T_S). The reason this does not occur in glass lies in the fact that the molecular building blocks (SiO_4 tetrahedra in silicate glass, Fig. 2-1) are spatially cross-linked to one another. In order to form crystals, the bonds must first be broken in order that crystal nuclei may form. This can only occur at lower temperatures at which the viscosity of the melt impedes the restructuring of the molecules and thereby the growth of crystals. The inclination toward crystallization (a glass specialist speaks of *devitrification*) generally decreases with the increasing speed of the cooling within the critical temperature range below T_S and with the number of components. Therefore, it is also influenced by the composition. De-

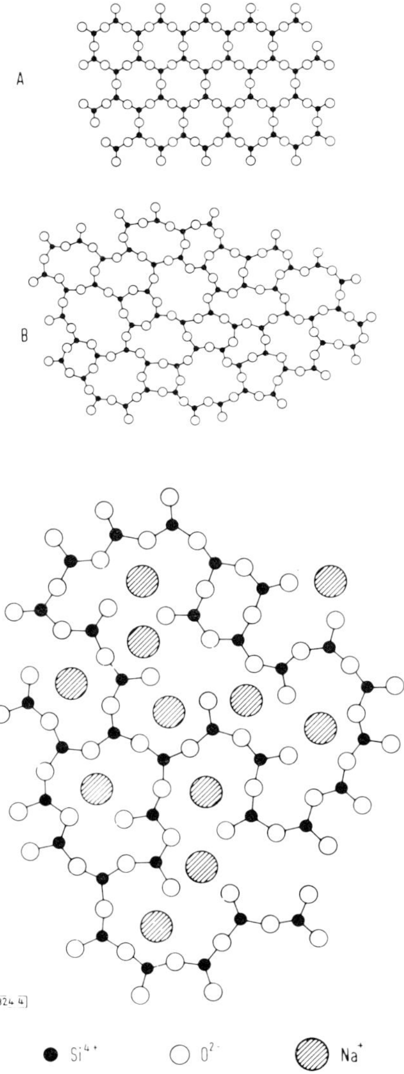

Fig. 2-1. Network of SiO_4 tetrahedra a) in crystal, b) in fused silica, c) in sodium silicate glass (two-dimensional display, the 4th oxygen bond of the tetrahedron is perpendicular to the plane of the picture).

vitrification is only desirable in glass ceramics. (See Glass Ceramics, Chap. 6.)

The difference between a system, which crystallizes when brought below the melting point and the same system, which, as a result of the hindering of crystallization (by rapid cooling, for example), hardens as glass, is especially clear if the volume is measured as the temperature falls. The result is shown schematically in Fig. 2-2; the temperature (T) is entered on the horizontal axis, the volume on the vertical axis. As soon as the temperature falls to the melting point (T_S), System 1 jumps from A to B during which time it forms a crystalline mass. On the other hand, System 2, as a supercooled liquid, condenses further to point C; and if the cooling is sufficiently slow, it will continue to condense further to point D. Here, the curve flattens out towards E or F, but it always stays above the base line B-G of System 1. So at room temperature (T_R), the system does not reach the density of the crystallized system.

Within the temperature range from point C to D, which is called the transformation temperature (T_g), the supercooled glass transforms from a plastic state to a rigid state typical for glass. The movement of the structural elements is very slight here, which is also demonstrated by the high viscosity of glass in this state. (It is possible, for example, to determine the viscosity from the speed by which a horizontal glass

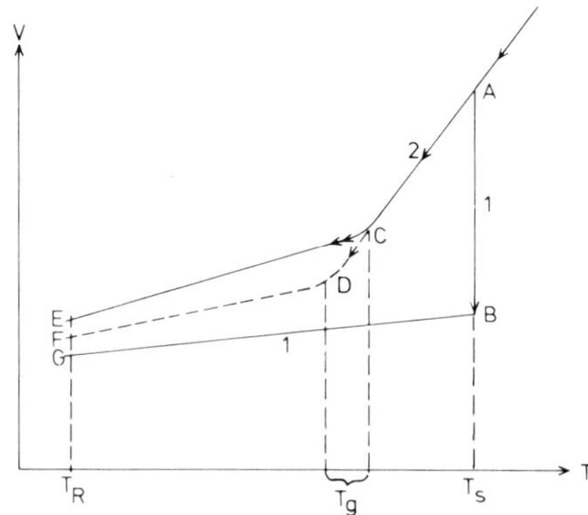

Fig. 2-2. Change in volume of a melt during (1) crystallization, (2) glass formation.

rod supported at the ends is bent by a load placed at its center.) This demonstrates that not only all inorganic glasses, but also all substances have a viscosity value of $\eta \approx 10^{13}$ Poise (P)* if they exhibit the characteristics of System 1 in Figure 2-2. Compare: At 20°C, water has a viscosity of 10^{-2} P, olive oil about 10^2 P, honey about 10^4 P.

The relationship of the viscosity of glass with temperature change (Fig. 2-3) is of basic importance in all facets of glass technology. It must be heated to a temperature at which $\eta \approx 10^2$ P in order to achieve a homogeneous melt. According to the process used, the glass formation takes place at 10^3 to 10^8 P. If the temperature range between viscosity values 10^4 and 10^8 P is large, such glasses are called *long glasses*; if it is small, they are called *short glasses*. These differences are very important for processing because of the deformation period available.

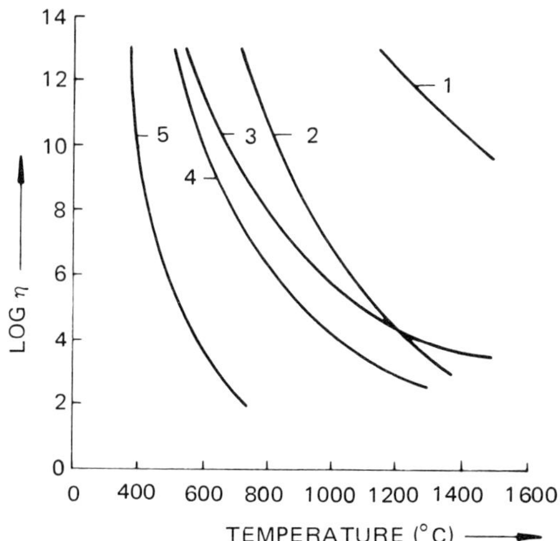

Fig. 2-3. Temperature dependence of viscosity (η) of several technical glasses. 1: Fused silica, 2: Alumino-silicate glass 8409 (see Table 6-1), 3: Borosilicate glass "Duran" (p. 111), 4: Soda-lime glass (p. 56), 5: Lead-borate soldering glass.

*In the new SI units, the poise units have been replaced by dPas (deci-Pascal seconds), which are equal to 0.1 N.s/m² (force unit N = Newton, corresponding to a weight of 102 grams). The old unit, P still appears frequently in glass literature. (See Appendix.)

In order to be able to distinguish the viscosity behavior of the different types of glass, several fixed points have been established which describe the temperatures of certain viscosity values. They are presented in the chart below.

10^4 P: Working Point (V_A)
$10^{7.6}$ P: Softening Point (E_W)
  Glass deforms due to the pressure of its own weight
10^{13} P: Annealing Point $\}$
$10^{14.5}$ P: Strain Point $\quad$ within the T_g area

For example, the fixed point temperatures of flat glass are illustrated on page 56, and those of several technical glasses are listed in the tables on pages 110 and 111.

The transformation from the plastic to the brittle state occurs in the transformation range. In the same region, mechanical stresses are eliminated which appear due to rapid cooling during processing. At 10^{13} P, 15 minutes are sufficient to accomplish this. At $10^{14.5}$ P, the elimination of stress can take many hours. Stresses in the glass may be located and measured by the optical birefringence that they cause.

The mechanical properties of glasses are also somewhat peculiar. Since the chemical bonding in the glass network and the energies required to break them are known, it is possible to derive theoretical tensile strengths. This results in values of the order of 10^4 N/mm². However, the technical strength is several hundred times less and is heavily dependent on the surface state of the glass. So, for example, the maximum permissible continuous load for architectural glass is only set at 7 N/mm². If the glass is subjected to suitable surface treatment (such as fire-polishing, protective coating, and prestressing— see Safety Glass, Chap. 4) to ensure that damage and microcracks are minimized, stability values of 5×10^2 N/mm² can be attained, which is still substantially below theoretical values. This occurs because internal defects and flow points are frozen within the glass during cooling. They limit the strength of the glass. The strength of thin, freshly drawn glass fibers is much higher because the number of flaws is low. (See Glass Fiber, Chap. 6.)

Experience with household glassware demonstrates that glass will break easily if its temperature changes rapidly, especially if the temperature changes from hot to cold. Several properties lead to this behavior: poor heat conductivity, the relatively high thermal expansion of alkali-rich glasses, and limited tensile strength. When a glass is chilled (or quenched) after being heated to a point below its T_g temperature, only the outer layer begins to cool, and it seeks to decrease its volume. But it is stretched by the core which is still hot and therefore subject to high tensile stress. If the tensile strength, reduced by surface scratches, is exceeded, fracture sets in. It starts at the external surface and rapidly travels inward. On the other hand, rapid heating is less dangerous since the outer layer comes under compressive stress, and the resistance to compression of glasses is at least ten times that of its tensile strength.

Entirely different conditions are present when rapid cooling occurs above the T_g temperature. Compressive stress appears in the surface of the glass increasing its tensile strength. (See Safety Glass, Chap. 4.)

BROAD CLASSIFICATION OF GLASS TYPES

The large number of glass types can be classified in several ways, for example, by chemical composition, use in the manufacture of glass products, or processing behavior. The most widely used classification is by chemical composition which leads to three chief groups: soda-lime glass, lead glass, and borosilicate glass. Glasses in these categories account for at least 95% of all glass types. The remaining 5% are special glasses manufactured for the most part in very small quantities. Thousands of special glass types have been developed, and many of them have special applications. They are discussed separately in Chapter 6.

With very few exceptions, most glasses are silicate based glasses, the chief component of which is silicon dioxide (SiO_2).

Soda-Lime Glasses

By far the greatest number of industrially produced glasses belong to a related family collectively called the soda-lime glasses. As the name indicates, soda and lime play a major role along with the main component, sand.

A typical soda-lime glass is composed of 71%-75% sand (SiO_2), 12%-16% soda (sodium oxide as the raw material soda ash or sodium carbonate), 10%-15% lime (calcium oxide from the raw material limestone or calcium carbonate), and the remaining materials for specific properties such as coloring. Sometimes magnesium replaces a portion of the calcium contained in the limestone, or potassium replaces the sodium in the soda. Even still, these glasses are similar and may be classified as soda-lime glasses.

Soda-lime glass is chiefly used for bottles, jars, everyday drinking glasses, and window glass.

The chemical and physical properties of soda-lime glass create the conditions for its wide use. Among the most important is its light transmission which makes it suitable for use as flat glass in windows. An additional advantage is its smooth, nonporous surface which allows bottles and packaging glass made from it to be easily cleaned. Soda-lime glass containers filled with drinks and foods do not affect the taste of the contents, nor do they contain any harmful substances. Their durability in aqueous solutions is sufficient to withstand repeated boiling (preserving jars) without introducing negative surface changes.

Indeed, the relatively high alkali content of the glass lowers the melting point as opposed to pure SiO_2 glass but also causes an increase in the thermal expansion coefficient (α) by about 20 times from $\sim 0.5 \times 10^{-6}$/K to 9×10^{-6}/K*. (See Fused Silica, Chap. 6 and Table on page 56.)

With reference to the meaning of the value α, all glasses have been divided into two categories: glasses with α-values below 6×10^{-6}/K are called *hard glasses*; those with higher α-values are called *soft glasses*.

Due to its high thermal expansion, the resistance of soda-lime glass to sudden temperature change is comparatively poor (See General Characteristics, Chap. 2). Therefore, cautious handling is desirable, for example, when filling with hot liquids.

---

*K (Kelvin) corresponds to the old unit, °C when related to temperature differences.

Lead Glasses

If lead oxide replaces much of the lime in the batch, the result is a glass type popularly known as *lead crystal*. Such a glass is composed of 54%-65% SiO_2, 18%-38% lead oxide (PbO), 13%-15% soda (Na_2O) or potash (K_2O), and several other oxides. Glasses with less lead content (less than 18% PbO) are called *crystal glass*. Differing amounts of the oxides of barium, zinc, and potassium can be added to the composition to replace partially some of the lead oxide. To protect the consumer from confusing names, German legislation restricts the use of such terms as lead crystal, crystal glass, etc. for glass articles used on the table and in the household. (See Glass Tableware, Chap. 5.)

Glasses containing lead exhibit a high refractive index and are especially suited for decorating by grinding. Their specific gravity is higher than that of soda-lime glass. In our daily lives, we usually see them as drinking glasses, vases, bowls, ashtrays, or as decorative items.

Borosilicate Glasses

Silicate glasses containing boric oxide comprise the third group, borosilicate glass. These glasses have a higher percentage of SiO_2 (70%-80%) than the previous two groups. The rest of the composition is as follows: 7%-13% boric oxide (B_2O_3), 4%-8% Na_2O and K_2O, and 2%-7% aluminum oxide (Al_2O_3).

Glasses having such a composition possess a high resistance to chemical corrosion and temperature change. For this reason, they are used in process plants in the chemical industry, in laboratories, as ampoules and medicine bottles in the pharmaceutical industry, and as envelopes for high-intensity lamps. But borosilicate glasses are also used in the home; baking and casserole dishes and other heat-resisting items can be made from it.

The family of borosilicate glasses is extraordinarily broad, depending upon how the boron compounds within the glassmelt interact with the other constituents. For this reason, most of these glasses are designated as special glasses. (See Chap. 6.)

Special Glasses

Glasses used for special technical and scientific purposes form a mixed group. Their compositions differ greatly and involve numerous chemical elements. This group includes the optical glasses, glass for electrotechnology and electronics, and glass ceramics. These are discussed in more detail in Chapter 6.

RAW MATERIALS FOR THE MANUFACTURE OF GLASS

Sand is the most important raw material for glass. Almost half the earth's surface consists of silicon dioxide (SiO_2), a main component of different sands and rocks. However, most sands do not possess the purity necessary for the production of glass, because they have large amounts of coloring oxides, especially iron oxide. Sand containing as little as 0.1% Fe_2O_3 is useless for making such products as plate glass because it gives the glass a greenish tint. The occurrence of sands containing 0.01%-0.03% Fe_2O_3 for the production of special technical glasses is very rare. (See Fig. 2-4.)

Fig. 2-4. Qualitatively valuable sand as a raw material for the manufacture of glass.

Iron is an even greater problem in the quartz sand required for the melting of optical glasses. The iron oxide content must be less than 0.001% (and sometimes only a small fraction of this), so low in fact that the content is expressed in parts per million (ppm), where 1 ppm = 10^{-4}%. The content of other coloring oxides, such as chromium, nickel, cobalt, and other undesirable impurities in optical quartz sand must be substantially less than this. There are only a few deposits scattered over the earth's surface which meet these requirements. For these stringent demands, the crushed material is subjected to an additional chemical cleansing, using suitable acids at high temperatures. The grain size of sand ideally measures between 0.1 and 0.4 mm.

In order to lower the melting temperature of sand (over 1700°C) and to be able to use appropriate melting containers, a fluxing agent, usually sodium oxide, is required. This is normally added in the form of a carbonate (soda ash), or sometimes a nitrate or sulfate. Such an alkali is seldom found naturally. Most of the alkali metals are bonded to halogens, mostly as sodium chloride (table salt). Thus glass manufacturing on an industrial scale could really only take place after it became technologically possible to transform alkali halides into oxide-bonded sodium. This was first made possible by the Le Blanc method, and later by the Solvay method for commercial soda (sodium carbonate) manufacture.

Soda Ash

Sodium carbonate (Na_2CO_3) is introduced to the glass batch as soda ash (an anhydrous, white powder). During the melting, the soda (sodium oxide) becomes part of the glass; the carbon dioxide is freed and escapes up the chimney.

Glauber's Salt

Sodium sulfate (Na_2SO_4) (discovered in the seventeenth century by the doctor and chemist Johann Rudolph Glauber for medicinal use) can, in its anhydrous form, be mixed with pulverized coal and added to the batch instead of soda ash. The sulfurous acid is freed, and the sodium oxide becomes part of the glass as in the case of soda ash.

Potash

Potassium carbonate (K_2CO_3) is a grainy, white powder formerly obtained by leaching wood ashes (mostly beech and oak) in large containers (pots). Today, it is made industrially from potassium sulfate. The potash decomposes into potassium oxide, which goes into the glass, and carbon dioxide, which escapes into the atmosphere. In the absence of coloring metals, potash yields a pure, colorless glass.

Stabilizers for Increasing Durability, Strength, and Hardness

There are a number of oxides of multivalent metals which when added to the batch lend the finished glass important physical and chemical properties essential to its usefulness. Some substances can reinforce the structural network resulting in improved chemical durability and mechanical properties. To this effect, the oxides of calcium (CaO), magnesium (MgO), aluminum (Al_2O_3), and zinc (ZnO), as well as boron trioxide (B_2O_3) play an important role. The addition of K_2O in place of Na_2O also usually makes glass more chemically resistant.

Lime

Calcium carbonate ($CaCO_3$) is found naturally as limestone, marble, or chalk. At a temperature of about 1000°C, the carbon dioxide escapes from the lime. Only the calcium oxide (or calcined lime) remains, and it enters the glass structure. Lime is added to the batch to improve the hardness and chemical durability of the glass. In flat glass, the lime is partially replaced by magnesium oxide which is found combined with lime in the raw material dolomite ($CaCO_3$ + $MgCO_3$). It lowers the melting temperature.

Alumina

Aluminum oxide (Al_2O_3) is usually added to the glass batch in the form of alkali containing the felspars abundant in many places (for example, $NaAlSi_3O_8$). The trivalent aluminum forms AlO_4-groupings in the glass which in the presence of an alkali ion arrange themselves in the network of the SiO_4 tetrahedrons, thereby filling gaps. This leads to improved chemical durability and increased viscosity in the lower temperature ranges.

Lead Oxides

The oxides, PbO (yellow lead) and Pb_3O_4 (red lead) are used to introduce lead into the glass. However, it is always present in glass as bivalent Pb^{2+}. Moderate additions of PbO into glass increase chemical durability. High lead content lowers the melting temperature and results in decreased hardness, but increases refractive index of the glass important for its "brilliance." (See Glass Tableware, Chap. 5.)

Barium Oxide

Barium oxide from $BaCO_3$ (witherite) is chiefly used in optical glass and crystal glass instead of lime or red lead. Glass containing barium is not quite as heavy as lead crystal, but achieves similar brilliance due to its high refractive index.

Boron Compounds

Boron tri-oxide (B_2O_3, the anhydride of boric acid, H_3BO_3), which is so important in special glasses, is found naturally in very few places. Sodium and calcium borates occur much more frequently however. Usually, these compounds must be chemically converted to pure boric acid, especially for optical glass.

Coloring Agents

Only pure chemicals are used as glass coloring agents. Different colors can be obtained by adding oxides of the so-called transition elements (copper, chromium, manganese, iron, cobalt, nickel, vanadium, titanium) or of the rare earths (primarily neodymium and praseodymium) to suitable base glassmelts. The colors yielded by these metallic ions are listed in the table below.

Copper	(Cu^{2+}):	light blue
Chromium	(Cr^{3+}):	green; (Cr^{6+}): yellow
Manganese	(Mn^{3+}):	violet
Iron	(Fe^{3+}):	yellowish-brown; see also "carbon yellow" coloring, (Container Glass, Chap. 4)
	(Fe^{2+}):	bluish-green
Cobalt	(CO^{2+}):	intense blue, in borate glasses, pink; (Co^{3+}): green
Nickel	(Ni^{2+}):	greyish-brown, yellow, green, blue to violet, according to the glass matrix
Vanadium	(V^{3+}):	green in silicate glass; brown in borate glass
Titanium	(Ti^{3+}):	violet (melting under reducing conditions)
Neodymium	(Nd^{3+}):	reddish-violet
Praseodymium	(Pr^{3+}):	light green

Intensive yellow, orange, and red colors are produced by the precipitation of colloids of precious metals as well as of selenium, cadmium sulfide, and cadmium selenide during the cooling of the melt. Secondary heat treatments are also used to obtain these colors (*struck glasses*, the best-known of which is gold-ruby glass). Brown and gray colors are obtained with combinations of oxides of Mn, Fe, Ni, and Co, which in high concentrations can also yield black. Undesirable color streaks can be avoided where necessary through use of oxygen-releasing substances such as refining agents. (See Melting Process, Chap. 3.)

At 400°-600° C, the surface of colorless glass can also be stained from yellow to reddish-brown. Silver stain is especially popular. (See Finishing, Chap. 5.)

Opacifiers

By mixing fluorine containing materials, such as fluorspar (CaF_2) or preferably today phosphates, small crystalline particles form in the glass, rendering the glass cloudy and opaque. Such glasses are used in opal table glass, in opaque or milk glass for architecture, and in flashed opal glass globes in the lighting industry. (See Chapters 4 and 5.)

Glass Cullet

Although glass cullet is not technically a raw material, it may be added to the melt. Each glass factory saves its cullet, which is in the form of pieces cut from flat glass or rejects and breakages from hollow glass. (See Fig. 2-5.) If the cullet stock is insufficient, then it can be "produced" or purchased. Cullet acts as a fluxing agent and promotes the melting of the sand. This conserves both energy and raw materials. The recycling of cullet is gaining popularity, under the influence of increasing energy costs and growing environmental consciousness. (See Fig. 2-6.) If a glass batch for container glass, for example, contains 25% cullet, it requires 5% less energy to melt it. Cullet comprises about 50% of the batch material for colored container glass (bottles). In some cases, cullet of a different manufacture can also be used without negative effect on the quality of the glass.

Fig. 2-5. 30%-40% cullet in the batch allows the greatest furnace efficiency providing the glass is of sufficient quality.

Fig. 2-6. Glass recycling container.

The Batch

The mixture of the individual raw materials combined in certain amounts to yield the desired glass type is called the *glass charge*. When the glass charge is carefully put together, mixed, and ready for melting, it is called a *batch*. (See Fig. 2-7.)

All raw materials are intensively tested for quality when they enter the glass plant. Based on the results, the batch house calculates any correction factors in order to compensate for composition variations of the raw materials. With the aid of electronic programming, batch composition can be fully automated in modern plants. To prevent the mixed ingredients from segregating prior to melting batches are prepared with a water content of 2%-4%.

Fig. 2-7. A batch house.

3. The Glassmelt

Melting is the central phase in the production of glass. The individual raw materials combine at high temperatures to form molten glass. The quality of the batch material, the type of heating energy, and the type of melting process used are determined by the glass type to be melted and the product to be made. Removal of the glass from the furnace for shaping and annealing are subsequent stages to the melting process.

MELTING FURNACES AND MELTING TANKS

Originally, only pot furnaces were used. They could hold one or more containers to melt the batch material. From these, continuous melting tanks evolved as an integral part of the furnace. Both types are still frequently used side by side.

Pot Melts

A pot furnace consists of refractory brick for the inner walls, silica brick for the vaulted roof (crown), and insulating brick for the external walls. As a rule a pot furnace consists of a lower section for preheating the fuel gas and the upper furnace proper which takes the pots and serves as the melting chamber. (See Fig. 3-1.)

Today, pot furnaces are only used to manufacture mouth blown glass products and special glasses. There is room for six to twelve pots in which different glass types can be melted. (See Fig. 3-2.)

A pot is made of refractory clay (sometimes mixed with sillimanite "grog"). It has to be heated slowly to 1000° C before it can be used for melting glass. It resembles an upside-down, rimless hat open at the top. In the United States, covered pots are generally used. They have an opening at the top for feeding the batch and removal of the glass. Formerly, glass factories prepared their pots within their own production facility. Now they are chiefly supplied by companies specializing in ceramic pot production. The capacity of a pot can be as

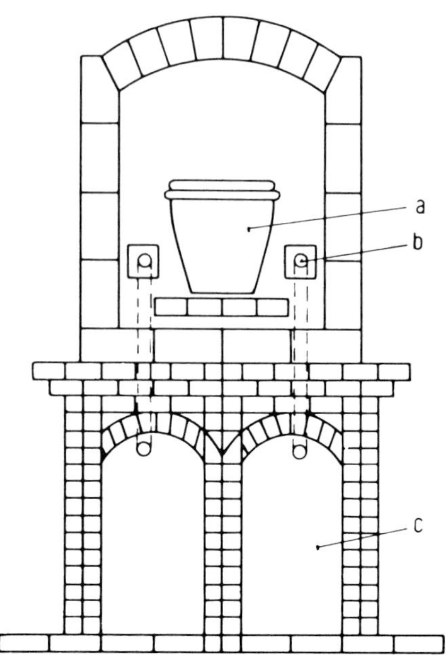

Fig. 3-1. Construction of a pot furnace (schematic) showing a pot and two regenerative chambers   a) glass pot   b) burner ports   c) regenerative chambers for heat recovery.

great as 2000 kg, but it is usually between 100 and 500 kg. The life of a pot averages eight weeks. (See Fig. 3-3.)

Pot furnaces are heated at all times, but their temperatures vary. The batch is charged in the pot in the afternoon, melted at night, and processed beginning early in the morning. During melting, the temperature climbs to 1300°-1600° C, depending on the glass type. During removal and processing of the glass, the temperature in the furnace is 900°-1200° C. A pot melt is not suitable for 24-hour production as the finished glass lasts normally for one shift only.

Tank Melting

For processing large amounts of glass, and above all, in automated production, tank furnaces are used. They consist of lower and upper sections as well as chambers in which the combustion air is preheated. The lower section is the actual melting container and consists of high-quality refractory material. The size and construction of tank

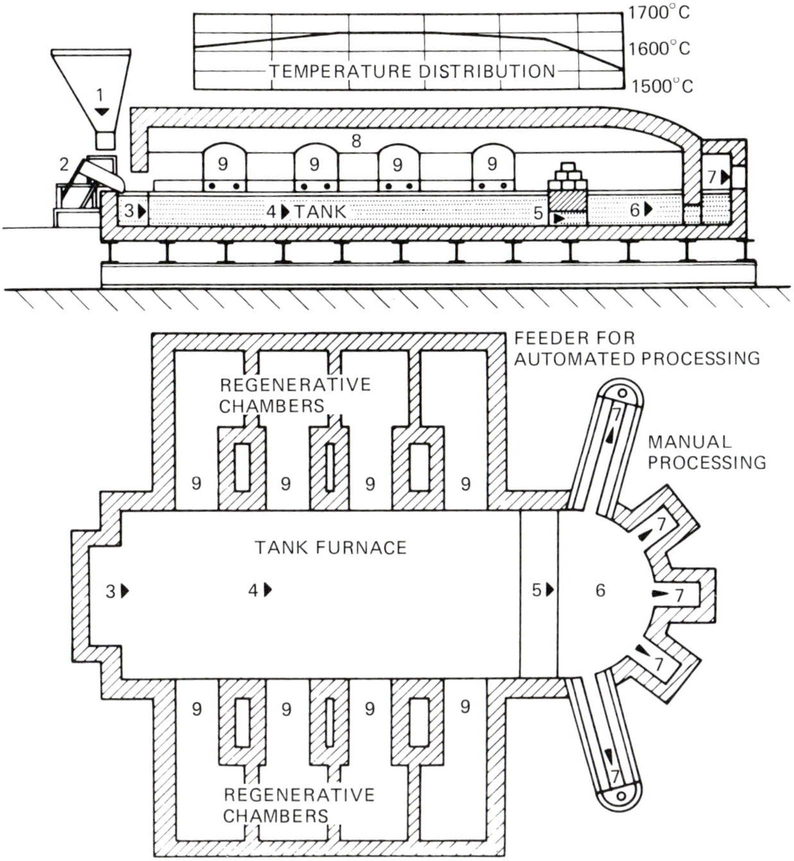

Fig. 3-2. Cross section and plan of a melting tank.

Fig. 3-3. Casting optical glass from a pot.

furnaces depends on the glass product manufactured. There are day tanks and continuous tanks. (See Fig. 3-4.)

Day Tanks.   These are a further development of the traditional pot furnaces and only have small capacities (such as, 10 tons per day). They are refilled with batch daily. The melting is done at night, and the glass goes into production the next day—indicative of their close technical relationship with pot furnaces. It is possible to change the glass type to be melted at short notice. This is known as changing melts. Day tanks are principally used for colored glass, crystal glass, and soft special glasses.

Continuous Tanks.   Their development was a prerequisite for the appearance of large glass companies with industrial mass production. They are exclusively used in the manufacture of flat glass, container glass, and certain mass produced optical glass types. Continuous tanks are 10-40 meters long and 3-6 meters wide. Their output lies between 100 and 400 tons daily, depending on the product. This re-

Fig. 3-4. Roof (or crown) of a melting tank.

quires continuous introduction of batch material and continuous processing of the melted glass. Tanks are nearly always operated on three shifts with fully automated production. Excluding the blast furnaces used in making iron and steel, glassmelting tanks are the largest industrial furnaces. The giants among them, used for making float glass, attain a length of 100 meters and a width of 13 meters. The melting tank itself can contain up to 2500 tons of molten glass. (See Fig. 3-5.)

Tank Construction

Continuous tanks differ chiefly in size and heating systems. The most common types are *regenerative side-burner tanks*, with preheating of fuel and combustion air; *regeneratively heated U-tanks for small output*, which require relatively little space; and *unit melters or long tanks*, a relatively recent design for bottle manufacture with horizontally opposed burner systems which save space, but use larger amounts of energy.

Fig. 3-5. Automated batch charging.

Materials for Furnace Construction

The refractory materials used for construction of glass furnaces, regardless of type, must be durable and economical. The selection of a refractory material for a specific use depends on its life, on the risk factor of contaminating the glass, and on the cost. Major developments in furnace construction material in the last four decades have permitted increased melting temperatures, improved melting efficiency, and better glass quality. The most critical furnace components are currently made of cast fused corundum-zirconium oxide refractory. The life of a tank furnace, called the *furnace journey*, is currently about eight years. The normal lifetime of a pot furnace is around five years. Within these periods it is possible to repair the furnace; these can be either hot (without ceasing the heating operation) or cold repairs (the furnace is shut down).

FUELS

As the chapter entitled "The History of Glass" has shown, wood was originally the only available fuel for melting glass. It was slowly replaced by coal beginning in the seventeenth century, but coal (both lignite and hard coal) was not in overall use until the beginning of the nineteenth century.

Gas

Like wood, coal was originally used to fire glass furnaces directly. However in 1860, Wilhelm Siemens discovered indirect heating by burning the gas which was obtained by the degasification of coal in a generator, the gas producer. This method was quickly adopted by the glass industry and producer gas was not replaced by more modern methods until after 1950. Then plants began to switch more and more from producing their own gas to purchasing it from a gas company. The quickly growing use of natural gas at the end of the sixties accelerated this trend. The advantages of modern gas firing are high purity, ease of control, and the absence of inventory costs.

Where glass plants are not located close to the gas grid, liquid gases (propane or butane) are used to heat small melting and processing facilities.

Fuel Oil

Fuel oil quickly became popular in the German glass industry after 1950. Light and heavy heating oils are distinguished by differing heating values. In 1975, the sources of energy in the German glass industry were distributed as follows: 37% gas, 52% fuel oil, and 11% electricity. The high amount of heating oil can be traced to its relatively low cost at the time, an advantage which was dependent on the pricing policies of oil producing countries and which has since vanished due to increasing world oil prices. One advantage of fuel oil over gas is the brighter burning flame which is, therefore, more easily adjusted; disadvantages are the high sulfur content of some fuel oils, the tendency to coke, and the necessity to maintain storage tanks.

Electricity

The use of electricity to melt glass goes back to around 1900. At that time, the glass bath was radiantly heated with heating rods; induction heating with middle-frequency alternating current and resistance heating of the containers were also in use. Heating the glassmelt with submerged electrodes is more effective and gaining increased popularity. In countries where electricity costs are low (Scandinavia), electric glassmelting enjoys wide use. In fully electrically heated tanks, a heating efficiency (expressed simply as the ratio of effectively used heat to supplied heat) of over 70% can be achieved; with flame heating (gas or fuel oil), 30% is seldom exceeded. In addition, electricity will not harm the environment, adds no impurities to the glass, and requires limited investment and repair costs. Electric melting of glass is by far most economical in smaller tanks which use less primary energy than comparable flame heated tanks. Electricity prices will finally decide whether this practice will continue to spread. (See Fig. 3-6.)

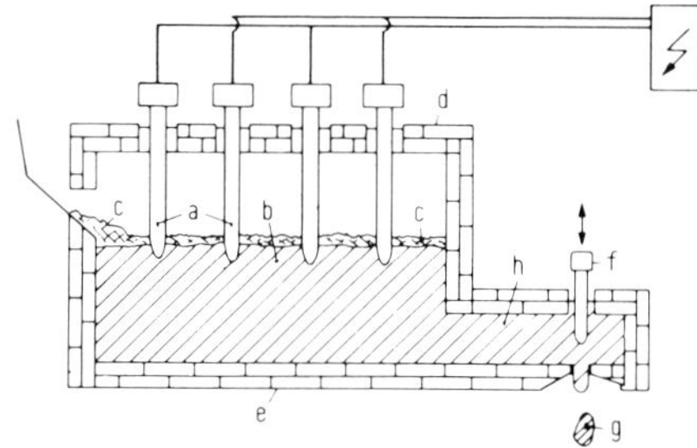

Fig. 3-6. Construction of an electrically heated glass tank (schematic). a) platinum electrodes, b) glassmelt, c) batch, d) crown, e) tank floor, f) plunger, g) measured gob, h) forehearth.

Heating

In heating the furnaces with gas or fuel oil, the combustion gases (gas or air) are either both preheated or the air only is preheated.

Regenerative Heating. Gas (piped gas, natural gas or producer gas), which is directed into the melting furnace along with cold combustion air, reaches a working temperature of 1000° C. Since temperatures of 1300°-1600° C are necessary for melting glass, the gas and air must be heated prior to mixing in so-called preheating chambers. The chambers consist of open brickwork through which the hot exhaust gases of the melting furnace are directed. By changing the flow at half-hour intervals, cold gas and cold air are heated in the chambers. This is the operation of regenerative furnaces using both gas and fuel oil.

Recuperative Heating. In recuperative heating, there is no exchanging of cold gas and hot exhaust gas within the furnace system. The direction of flow of the combustion gases (gas or heating oil) and the exhaust gases is always the same. The exchange of heat is accomplished by means of a thin dividing wall between the cold and hot gases which facilitates heat exchange.

THE MELTING PROCESS

The melting process is divided into several phases which require close supervision and control. This holds true for the traditional pot furnaces as well as the most modern tank furnaces.

Primary Melting

Due to the poor thermal conductivity of the batch, the temperature development in a newly added batch is so slow that there is time for different physical and chemical processes and reactions to occur between the separate ingredients of the batch. Some of the raw materials decompose under the effect of the heat, gases combined in the raw material escape, and the moisture in the batch evaporates. The raw materials slowly melt between 1000° C and 1200° C. (See Fig. 3-7.) First, the sand begins to dissolve under the influence of the fluxing agents. The silicic acid from the sand combines with the sodium oxide in the soda ash (or the potassium oxide, depending on the fluxing agent) and with other glass modifying batch materials. At the same time, large amounts of gases escape through the decomposition of the hydrates, carbonates, nitrates, and sulfates, in the form of water, carbon dioxide, nitrogen, oxygen, and sulfur dioxide. For example, one liter of soda-lime glass batch gives off about 1440 liters of gas at 1000° C, 70% of which is carbon dioxide (CO_2). In reality, the glass-

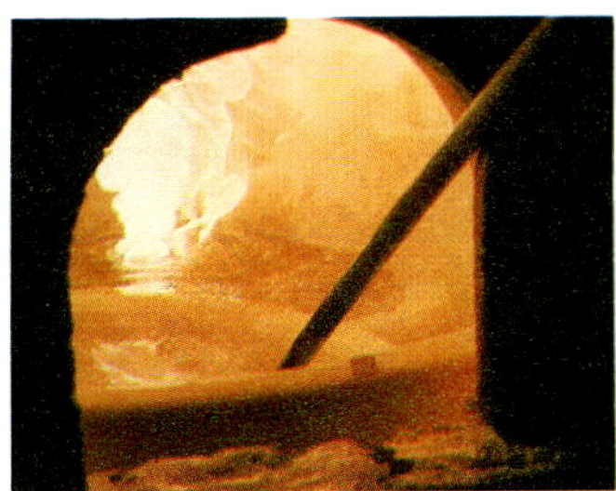

Fig. 3-7. Glass melting in a pot at ca. 1200° C.

melting process is even more complicated. The glassmelt finally becomes transparent, and the melting phase is completed. The volume of the molten mass has decreased to about a third, since the small, empty spaces between the grains of batch material have disappeared, and all gases have escaped. In pot melting, more batch material must be added in order to get the required amount of glassmelt. In tank furnaces, batch material is usually continuously added by automatic batch chargers.

Refining

While the batch is melting, the long process of homogenization begins. Homogenization involves the complete dissolution and even distribution of all components (especially important for the elimination of striae), as well as the refining process which eliminates all bubbles from the melt. In order to achieve the maximum homogeneity and freedom from bubble, a thorough mixing and degasification of the glass is necessary. Chemical refining agents are most frequently used to accomplish this on the principle that the added compound which gives off gas at high temperatures (for example, arsenic pentoxide [As_2O_5]) decomposes into arsenic trioxide (As_2O_3) and oxygen at approximately 1250° C. Sodium sulfate (Glauber's salt) has a similar effect, in that it gives off sulfur dioxide and oxygen at approximately 1200° C. It is preferred as a refining agent for mass-produced glass because it is the least expensive. The oxygen bubbles absorb the other dissolved gases and bubbles, thereby enlarging them so that they rise faster to the surface. Refining is also achieved by bubbling processes. Steam, oxygen, nitrogen, or air is forced through openings in the bottom of the tank. By increasing the temperature to give lower glass viscosity, the gas bubbles can rise more easily to the surface. This is known as *plaining*. Stirring mechanisms are used in the melting of optical glass in order to attain the high degree of homogeneity required.

Conditioning

A conditioning phase at lower temperatures follows the melting and refining stages. During this process, any remaining soluble bubbles are reabsorbed. The dissolution of any remaining bubbles is an important part of refining. At the same time, the melt cools slowly to a working temperature between 900°-1200° C.

These steps occur sequentially in pot melting. In tank furnaces, which are in continuous operation, the melting phases occur at different locations within the tank. In other words, the batch is fed at one end of the tank and flows through different zones in the tank where primary melting, refining, and conditioning occur, until the working temperature is reached at the other end of the tank. The refining process in a tank is actually more delicate. Considering the importance of tank melting for the industrial manufacture of glass, it is described separately.

Refining in a Tank Furnace

Glass does not flow through the tank in a straight line from the hot end where the batch is fed to the cold end where the glass reaches the working temperature for processing, but instead it is diverted due to thermal currents. The batch pile, or the cold mixture of raw materials, is not melted at the surface, but on the underside by the molten glass bath. Relatively cold, bubbly glass forms below the bottom layer of batch material and sinks to the bottom of the tank. If it did not reach the surface again due to thermal currents, it would not be refined since refining occurs in tank furnaces mainly at the top of the tank, where bubbles need rise only a short distance to escape. Furthermore only there are the temperatures high enough to cause bubble growth sufficient for refining. If thermal currents flow too fast they hinder refining by bringing the glass to the conditioning zone too soon. By introducing structures in the tank which obstruct the glass flow (e.g., a bridge-wall) ideal current paths can be induced.

Heat Required in Glass Melting

The *theoretical heat requirement* is that amount of heat energy needed to melt the glass and achieve the required end temperature. It is composed of the melting heat (the heat required for glass formation from the batch), and the retained heat of the glass and gases formed, up to reaching the end temperature of the glass. To assess the efficiency of tank furnaces, information is required on furnace idle values, daily throughput, glass take-off conditions, daily heat energy consumption, and heat transfer efficiency. The *furnace idle value* is the heat required to maintain furnace temperature with no glass take-off. *Heat transfer efficiency* is the ratio of useful heat energy (equal to melting capacity multiplied by theoretical heat requirement) to actual heat energy consumption.

Batch Charging

In pot furnaces the batch material is added by shovel. (See Fig. 3-8.) When the furnace is filled with molten, refined, and conditioned glass, the glass is gathered and processed. Only then is more batch material added. This is a 24-hour cycle. It is done differently in most modern tank furnaces which are usually constructed with a batch charging compartment (the "doghouse"). The molten glass surface in the doghouse is covered with batch material and the floating batch is pushed away. The advantages of feeding the furnace through the doghouse lie in the reduced amount of dust emitted and consequently longer furnace life spans, less radiant heat loss through the charge opening, ease of operation, and direct connection to the batch silo.

Automatic ram-feeding equipment has been developed recently. In this method batch charging machines cover a large part of the melt surface with an unbroken layer of batch having a depth of 3 to 8 cm and almost reaching the tank walls. This has a beneficial effect on throughput capacity. For the same temperature and glass quality con-

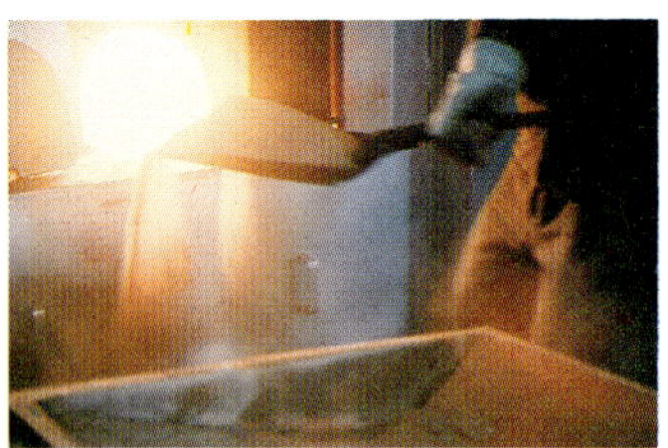

Fig. 3-8. Charging a pot furnace.

siderable increases in throughput can be achieved. The most important contribution of the mechanical batch charger to glass processing in tank furnaces is the maintenance of a constant glass level. This is automatically measured by mechanical, electrical and optical equipment, and controlled by adjusting the batch charge rate.

Melting Defects

There are a number of melting defects visible in the finished product which have a negative effect on its usefulness. They are of equal importance to both glass producer and user. (See Fig. 3-9.)

Stones. These are nontransparent grains of various sizes. Devitrification stones, as they are called, appear because molten glass has crystallized. Stones can also be small particles of the refractory material which have broken off the tank or pot walls. Batch and melt stones result from raw material grain sizes being too large, or insufficient melting temperature.

Striae (Cord). These are streaks of inhomogeneous, transparent glass in glass. They are caused by the batch or particles of refractory material, and can also arise on the glass surface during processing. Optically, striae are zones of glass of different refractive index and produce distortion of light transmission visible in polished articles. Very intense striae are known as cords and have the appearance of ropes or strands within the bulk of a piece of glass. (See Fig. 3-10.)

Fig. 3-9. Taking a sample from a tank furnace.

Fig. 3-10. Striae in glass.

Bubbles.   These are mostly caused by insufficient refining of the melt. Large bubbles are referred to as *blisters* and small bubbles as *seeds*. Longitudinally stretched bubbles are often called *air-lines*.

Discolorations.   These are caused by impurities in the raw materials or insufficient decolorizing of the melt.

4. Flat Glass

The term *flat glass* pertains to all glasses produced in a flat form, regardless of the method of manufacture. Unlike hollowware, it was not until the Middle Ages that glassblowers were finally successful in making flat sheets of glass for windows and further processing. Mechanical production processes were developed much later, effectively after 1920 (after earlier beginnings). Today manual production of flat glasses is the exception rather than the rule.

Over one-third of the flat glass produced in Germany is not used in its original form, but finished into other products, such as automotive safety glass and mirrors.

By far the greatest amount of flat glass consists of soda-lime glass. All manufacturers use basically the same formula. Average values are: 72% SiO_2, 14% Na_2O (+ K_2O), 9% CaO, 3%-4% MgO, and 1% Al_2O_3. The iron content of the raw materials gives thick soda-lime flat glasses a green tint. Variously tinted glasses for special purposes are also produced through the addition of coloring oxides to the same base glass mix. Borosilicate sheet glasses and "white" special mirror glasses, on the other hand, are virtually colorless. (See Other Types of Flat Glass, Chap. 4.) Their iron content ($\leq 0.03\%$) is 5-10 times less than that of common soda-lime glass.

THE PRODUCTION AND USE OF COMMON TYPES OF FLAT GLASS

Rolled (or Cast) Glass

Rolled glass is a poured and rolled flat glass which is not totally transparent.

Production of Rolled Glass. The molten glass flows out of the melting tank over a refractory barrier called the *weir* onto the machine slab. From there, the glass enters between two water-cooled rollers. Their distance from each other determines the thickness of the glass plate. A refractory "gate" regulates the amount of emerging glass. If wire-reinforced glass is being made, wiring equipment is located in front of the forming rollers. Wire net rolling is introduced into the hot glass by means of a locating roller. If the cast glass is to be ornamented, that is, if the surface is to have a design, shaping rollers with the desired surface relief are used. They serve as a continuous embossing stamp. The application of the surface design precedes the annealing process. The glowing ribbon of glass is transported over rollers into a tunnellike annealing furnace or *lehr*. (See Fig. 4-1.) Inside, the cast glass is first reheated to 600°-800° C. On the way through the annealing tunnel, the temperature decreases slowly in a carefully calculated manner, so that no internal stresses arise within the glass. At the cold end, the finished product, nearly handwarm, is ready for use. It is cut to finished and standard sizes and packed in shipping crates. Handling and loading equipment have rubber suction pads to prevent breakage. Cutting operations can also be computer operated without human assistance.

Gas is usually used to heat the lehr, but sometimes fuel oil is used. It is usually constructed of brickwork encased in sheet metal. To accelerate cooling, the glass lies on an open track for the last section of its passage.

Fig. 4-1. Machine casting of flat glass.

Uses of Rolled Glass.   Rolled glass is used where clear visibility is either superfluous or not desired. It is translucent with a light transmission of 50%-80%, depending on the thickness and surface structure. It is most frequently used in such areas as bathrooms, washrooms, house and office doors; and glazing in industrial buildings, railroad stations, and elevator shafts. Rolled glass is also popular in canopies, partition walls, partition panels between counters in banks and government offices, and spandrel cladding. It is also suitable for use in tabletops and lights for indoor and outdoor use.

Types of Rolled Glass.   Rolled glass is categorized, according to its appearance, as raw glass, ornamental glass, greenhouse glass, wire reinforced ornamental glass, and profile glass. There is also colored rolled glass.

Raw Glass   is a rolled glass with the surface rolled smooth or slightly figured. It can be used in applications which do not require transparency or specific optical properties. Raw glass can have a "hammered" (like metal), coarse-ribbed, or faceted surface. Such configurations appear on one surface only. The other is *fire-polished*, that is, cooled down immediately after being flat rolled.

Ornamental Glass   is rolled glass with the surface figured by shaping or embossing rollers. A large variety of designs can be given to one or both sides. The designs are not only decorative. Surfaces are often specifically designed for high dispersion and reduction of glare. Glass manufacturers generally have their design names protected legally. Often the name indicates the type of ornamentation used: wave glass, light-scattering glass, roundel (or bullion or bull's-eye) glass, lined glass, or trade names like Karolith or Difulit. Some figured designs are shaped like flowers or snowflakes.

Greenhouse Glass   is a translucent rolled glass with a special surface design to scatter light. It is used in horticulture. In hothouses and garden frames, it serves to scatter incident sunlight and heat evenly. It should not be confused with horticultural clear glass which is a type of sheet glass manufactured by a different process.

Wire Reinforced Glass is made, as has been mentioned, by embedding wire net in rolled glass. The embedded wire material can be made of 6 mm mesh webbing, 12.5 mm mesh spot welded netting, or 25 mm mesh hexagonal netting. Common wire reinforced glass is a raw glass and therefore transparent. It serves to hinder break-in and fire. For this reason, it is generally found in house doors, hall doors, or storage area doors, where its use is officially accepted for fire-resistant construction elements. On breaking, the wire inlay holds the glass together to provide security. Smashed wire reinforced glass will not fall, thus reducing the risk of personal injury from accidents involving doors and elevator shafts. In roof glazing, particularly in factories, it can carry loads (fallen objects, icicles, snow, etc.) even after the glass is broken.

Wire Reinforced Ornamental Glass is similar in manufacture and application to plain wire reinforced glass. The difference lies in the ornamentation of one or both surfaces. The surface design is similar to that found in unreinforced ornamental glass. Wire reinforced ornamental glass has practical importance when both decorative effect and safety factors are of concern. It is usually used indoors.

Profile Glass is a U-shaped rolled glass for architectural use. Profile glass is manufactured in continuous bands of glass with both edges already bent upwards. The dimensions of the base are 23-26 cm, and the limb height is 4-6 cm. It is mounted in metal frames which have a corresponding U-shaped cross section. The joints are permanently, elastically sealed. It can be singly or doubly glazed, and is used for skylights and walls in factories, warehouses, staircases, or multi-story garages.

Colored Cast Glass includes many kinds of cast and rolled glass. There are more than 100 colors of *dalle* glass (*dalle* is French for "tile") to choose from. *Dalle* glass is produced in pot furnaces and hand cast (without rolling) into plates which are approximately 25 mm thick. These are used in special studios to make *concrete-glass windows*. The glass is broken to the desired size and arranged on a drawing according to an artistic design. The "composition" is then placed within a steel frame, and concrete is poured into the open spaces and allowed to harden. This type of window is popular in church construction. (See Fig. 4-2.) *Dalle* glass is also used to make door handles and all-glass doors. (See Fig. 4-3.)

Opaque Glass   (opaque, colored flat glass) transmits no light whatsoever. The front side is fire polished and the reverse side ribbed. Opaque glass can be cut, ground, drilled, and otherwise processed (etched, frosted, etc.). It usually comes in white, black, green, grey, and beige; there are also multicolored, veined opaque glasses.

Since opaque glass is also air- and water-tight and not transparent, it is used as wall-cladding for different exterior and interior architectural applications as well as for furniture and other equipment (shelves and display bases). As wall tiles, it is mounted with mortar

Fig. 4-3. Door handles made of *dalle* glass.

Fig. 4-2. Church window made of *dalle* glass.

or special adhesives. Large sizes are used as weather-resistant facade cladding of buildings. Opaque tiles or panels are resistant to ageing, since the hairline cracks which appear on ceramic cladded surfaces do not occur. Opaque glass is very suitable for hygenic applications, such as in operating rooms, test laboratories, and pharmaceutical manufacturing plants. It is relatively resistant to acids and alkalis and unaffected by climatic temperature changes and frost. Limits to its application are imposed because of the cost of the special mounting techniques.

Opaline glass is closely related to opaque glass. It is an opaque cast glass with ground and polished surfaces. It has many of the same uses as opaque glass. Opaline glass is preferred when high-quality surface finish is required. It is also used for panels and signs.

Window and Plate Glass

Only with the greatest difficulty and after years of experimentation were glass process engineers successful in drawing flat glass directly from the tank. Up until 1900, window glass was exclusively made by the glassblowers who blew the glass into a large cylinder which was then cut lengthwise, put into a special furnace called the flattening furnace, and processed into glass sheets by heating, stretching, "ironing-out" and finally annealing into panes of window glass. The first unsuccessful efforts to make window sheet glass by drawing or lifting molten glass directly from the furnace go back to the 1850's.

The Fourcault Process. In 1905, a Belgian named Fourcault succeeded for the first time in drawing sheet glass directly from the tank. His process went into industrial production in 1914. The most important part of the process is the *debiteuse*, a three meter long clay block with a slot which floats on the molten glass. If it is pushed slightly into the glass, the molten glass rises out of the slot and is grasped by an iron "bait." The thickness of the sheet is irregular initially, but evens after running. Rollers draw the hardened glass about 7 meters up the annealing shaft. When it emerges from the shaft it is cool, annealed and ready to cut. (See Fig. 4-4.) The slower the drawing

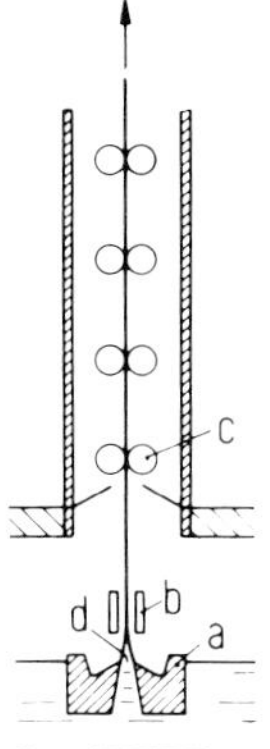

Fig. 4-4. The Fourcault method for drawing sheet glass. a) *debiteuse,* b) coolers, c) drawing rollers, d) dome.

rate, the thicker the glass. Up to seven *debiteuses* with annealing shafts can be operated in one Fourcault tank. Major disadvantages of the process are: (a) the fine roller marks on the surface of the glass, and (b) the tendency to devitrify due to the effect of the refractory material of which the *debiteuse* is made.

Libbey-Owens Process. Somewhat later than Fourcault, the American Colburn successfully devised another method of making window glass. With the support of the American firm Libbey-Owens, the process was further developed. It was given the firm's name when it went into industrial production in 1917. Unlike Fourcault's system, Colburn's process does not have a *debiteuse*. Initially the glass is drawn vertically from the tank by a metal "bait." It is immediately taken over by cooled side-rollers to prevent it from flowing together. After travelling about 70 cm, the still soft glass is led over a polished steel roller, and diverted into a horizontal direction before entering a 60 meter annealing lehr. The drawing speed is twice that of the Fourcault process. Libbey-Owens machines have two work channels from which glass emerges in a continuous band, as with the Fourcault process. (See Fig. 4-5.)

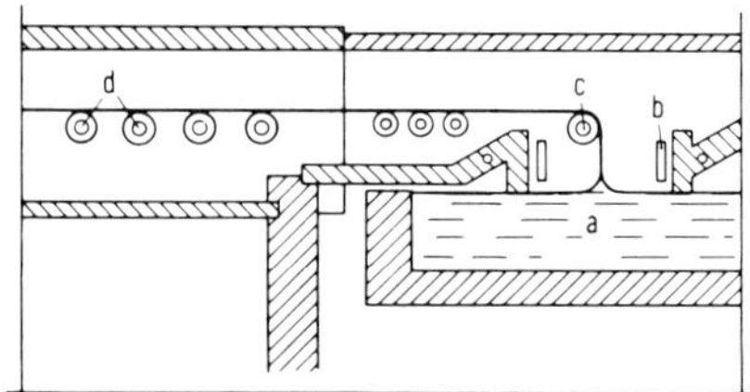

Fig. 4-5. The Libbey-Owens process for drawing sheet glass. a) glass bath, b) cooler boxes, c) bending roller, d) transporting rollers in the drawing and annealing channel.

The Pittsburgh Process. The production system developed by the American firm Pittsburgh Plate Glass Company combines the best features of the Fourcault and Libbey-Owens processes. It has been in use since 1928. In this process, the glass is first drawn perpendicularly like the Fourcault process. A *debiteuse* is not necessary, however. Instead, a guidance device made of refractory material is located in the glass at the drawing position. Cooled grippers, shaped like hollow plates, receive the glass. Their slit-shaped cutouts prevent the glass from binding. After passing through an annealing shaft about 12 meters long, the finished product is cut. (See Fig. 4-6.) The advantages

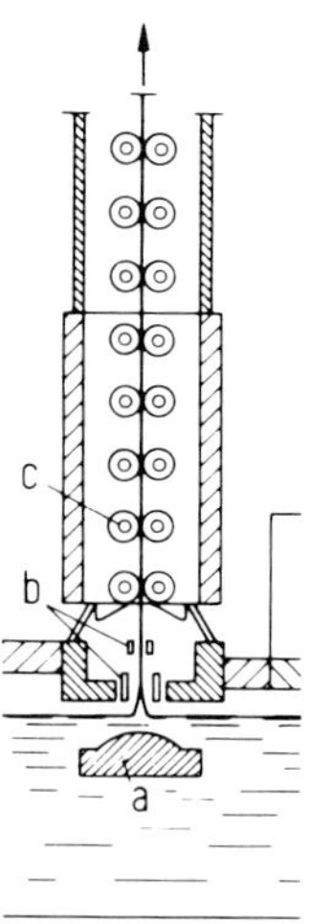

Fig. 4-6. The Pittsburgh process for drawing sheet glass. a) drawing rods, b) cooling zone, c) drawing rollers.

of the Pittsburgh process lie in its production speed, ability to change glass thickness quickly, and good quality.

The Uses of Sheet Glass.   Sheet, window or drawn glass (all names for the same product) always has two smooth or *fire polished* surfaces which require no further processing after emerging from the machine. It had been the most important type of flat glass for the construction industry over the years, but since the sixties, it has increasingly been replaced by float glass. Window glass is divided into different categories of thickness: *thin glass* (0.9-1.6 mm), *standard thickness* (1.8 mm), *medium thickness* (2.8 mm), *double thickness* (3.8 mm), and *thick glass* (4.5-6.5 mm). Thin glass is used in the glazing of pictures, as microscope slides, or as protective glass for instruments. Window glass in the next three thicknesses are primarily used in building construction. Thick glass was formerly used for such things as store windows and tabletops in furniture manufacturing, but very little is made today.

Poor quality window glass is used in greenhouses and cold frames (horticultural clear glass).

Plate Glass

For a long time, ground and polished plate glass was actually the highest quality flat glass. It is especially suitable for mirrors, since exceptionally distortion-free glass is necessary. Until the 1960s, plate glass was the predominant starting material for glass finishing operations in Germany and other countries with highly developed glass industries. The introduction of the float glass process nearly put an end to the old method of manufacturing plate glass, with the exception of "white" plate glass which is still manufactured in some special glass factories. (See Other Types of Flat Glass, below.)

To manufacture plate glass, rolled or thick glass was rough- and fine-ground with grinding compounds on large tables or conveyors equipped with large rotating disks. It was then polished smooth with polishing materials (iron oxide or cerium oxide). Later, large equipment was built in which grinding and polishing were done simultaneously on both sides in a continuous process (the *twin process*).

Float Glass

Since the beginning of the 1960s, the production of sheet and plate glass has been increasingly replaced by a modern process, which produces glass of practically plate glass quality, but like sheet glass is made in a single operation requiring no further processing. This was accomplished after years of research by the British firm Pilkington Brothers Ltd. in St. Helens, Lancashire.

The Float Process.   The verb *to float* means "to be bouyant." And this is basically the principle on which the process is based.

Glass types and melting methods are similar to those used in the other flat glasses. The difference is the molten glass is fed into a float bath of molten tin. This tin bath is 4-8 meters wide and up to 60 meters long. To guard against oxidation of the tin by atmospheric oxygen, the tin bath must be carefully maintained in a weakly reducing, carefully controlled protective gas. Since the glass thickness depends on the properties of the interface between glass and tin, which in turn depend on the composition of the atmosphere, the latter must be maintained exactly constant. (See Fig. 4-7.)

The glass floats like an endless ribbon on the tin. At the entrance, where the glass first makes contact with the tin surface, the tempera-

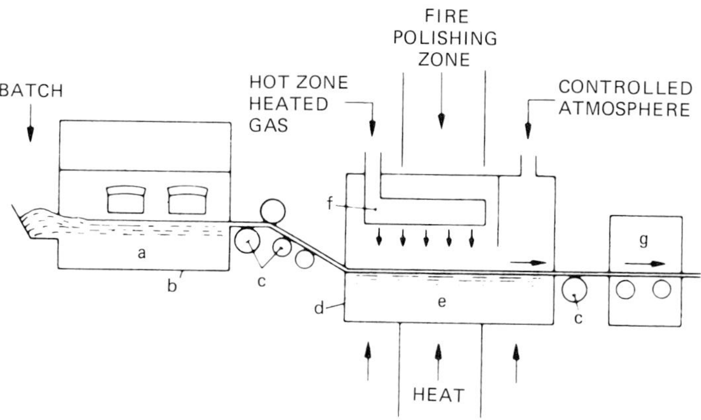

Fig. 4-7. Method of making float glass. a) glassmelt, b) glassmelting tank, c) transport rollers, d) float bath, e) molten tin, f) heated zone, g) annealing lehr.

ture of the molten metal is about 1000° C. At the exit, it is about 600° C. Tin is the only metal which is liquid at 600° C and does not develop appreciable vaporization pressure at 1000° C.

Just after the exit from the float chamber, special rollers take up the glass and feed it into the annealing lehr from which it exits at about 200° C. After cooling to room temperature on an open roller track, it is cut, packed, and stored either for later shipment or for processing into such products as safety glass or double- (or multi-) glazed units.

Fifteen years of experience have permitted thicknesses between 1.5 mm and 20 mm to be made. There are two techniques for doing this. To produce thin float glass, rollers are used to control the width and speed of the glass ribbon. For thick float glass, the glass floats against graphite barriers, so that the ribbon emerges thicker. In this manner the desired widths and thicknesses are achieved. In recent years colored float glass has also become available. The output of a fully automated plant can reach 3000 square meters per hour and requires only four people to operate it. In 1978, there were an estimated 50 float glass plants in operation; several more were also either already under construction or in the planning stages. The production of plate glass has been all but abandoned in Europe and has been replaced by the float process. Sheet glass is still drawn in several places, primarily using the Fourcault process which is the oldest fully automated method of producing flat glass. Most of these plants manufacture thin glass, since there are still technical difficulties in making float glass any thinner than 1.5 mm. (See Fig. 4-8.)

Fig. 4-8. Float glass plant of Flachglas AG in Weiherhammer, West Germany.

The Uses of Float Glass.   Since this new type of flat glass combines the qualities of plate glass and sheet glass, float glass is used for glazing wherever transparency is required in buildings. It is also a raw material for making safety glass, primarily for automotive and other transportation uses. Mirrors and otherwise finished or processed flat glass for furniture or interior decorating, are also made from it. In addition, there are numerous other uses in precision mechanics, especially where flat surface planes are required, as, for example, in visual displays.

TECHNICAL IDENTIFICATION OF SODA-LIME FLAT GLASSES

The following table contains important physical and technical information pertaining to the use of normal flat glass.

Density	2.5 g/cm³
Thermal expansion (linear)	$8.5\text{-}9.5 \cdot 10^{-6}$/K
Thermal conductivity (*k-value*) at 5mm thickness	5.6 W/m² K
Modulus of elasticity	$\sim 7.10^4$ N/mm²
Bending strength	~ 30 N/mm²
Specific electrical resistance	$10^{12}\text{-}10^{13}$ Ω cm
Transmission of light between the 0.36 to 2.5 μm wavelengths (light and near IR) at 4 mm thickness	$\sim 75\text{-}90\%$
Viscosity data* (see p. 19):	
Transformation temperature ($\sim 10^{13}$ P)	525-545° C
Softening point ($10^{7.6}$ P)	710-735° C
Working point (10^4 P)	1015-1045° C

Similar values hold true for common soda-lime container glass.

OTHER TYPES OF FLAT GLASS

In addition to rolled, sheet, plate, and float glass, there are a number of other flat glasses. Most of them have very specialized applications and are subject to stringent technical requirements. For this reason, they are generally considered to be special glasses, a subject to which a separate chapter is exclusively devoted. But there are several other

---

*See footnote, p. 19.

types of flat glass encountered in daily life which merit mention here. Most of these glasses are made by glass companies which belong to the Schott Group.

As already mentioned, near-colorless, temperature-resistant glass is obtained by using raw materials which contain the least possible amount of iron. Borosilicate flat glasses are especially important in the consumer markets under the designations *Tempax* (Schott) and *Pyrex* (Corning, U.S.A.). Tempax is manufactured by a special drawing process, while Pyrex is rolled. Panels of up to 2 square meters manufactured in these processes usually have fire polished surfaces. If flatter surfaces are required, the glass is then ground and polished conventionally.

Due to the high thermal resistance of these glasses (see Other Hollowware, Chap. 5 and Borosilicate Glass, Chap. 6), they are used as viewing windows in household stoves, cookers, washing machines, and the like. They are also used for colorless substrate glasses in hot panels with electrically conducting surface layers, and for heat filters and mirrors.

The flat glass made by Deutsche Spezialglas AG in Gruenenplan, West Germany merits mention as a completely colorless plate glass. This glass, identical in composition to ophthalmic glass, is processed like classical plate glass, i.e., it is ground and polished. Drawn sheets with 0.05-0.2 mm thicknesses are also made of it. Thin microscope cover slips are an example of the type of use to which this alkali-boro-zinc-silicate is put. These glasses resist thermal shock with the same effectiveness as common types of flat glass.

Antique Glass

Flat glass of the highest purity and clarity is not always desired. Antique glass has the deliberate appearance of pre-Industrial Age flat glass. (See Fig. 4-9.) The surface structure is uneven and there are internal bubbles caused by incomplete refining or by the addition of gas-producing substances to the glassmelt shortly before it emerges for processing. According to the method of production, antique glass can be classified as either a sheet or a plate glass. It can be made by

Fig. 4-9. Panels of antique glass.

the original (or antique) method from mouth blown cylinders or drawn by machine. If only one side of the glass has a surface structure, it is called *neo-antique glass*. These designations have been generally adopted, even though they are misleading. In a strict sense, none of this glass is antique or old. Rather they are reproductions of antique glasses. When antique glass is mechanically drawn, the surface of the plates show a repeated design called a *register*.

Antique glass is popular for decorating in the rustic or nostalgic style, in which case it is often glazed with lead or brass strips into furniture and windows. In living areas or inns decorated in the old German style, antique glass is regularly encountered.

So-called rolled antique glass is different from neo- or genuine-antique glass. It is an ornamented rolled glass with antique-glass-like surface structure.

Flashed Glass

Not all colored glasses have body coloring. When a colorless glass is surrounded by a layer of colored or opal glass, it is called *flashed glass*. The flashing technique is used for flat glass and hollow glass (lamp globes, tableware, and decorative glass). Otto Schott introduced it for technical use.

Flashed opal is the best known glass in this category. Instead of using the traditional mouth blowing techniques, there is now a mechanical process. In this process, an endless ribbon of colorless glass is drawn from a Fourcault tank through a *debiteuse* like sheet glass. At the same time, a thin ribbon of opal glass having similar thermal properties is drawn from a small tank located nearby. The fusing of both ribbons of glass takes place where the colorless glass emerges from the *debiteuse*. Glass thicknesses between 1.5 mm and 7 mm can be achieved with this method.

Flashed opal glass is well-suited for evenly dispersed, glare-free space illumination, for translucent room dividers, for bathroom windows, for shelves in showcases, X-ray picture viewers, or as indicator backs in technical and medical measuring instruments. (See Fig. 4-10.)

PROCESSED FLAT GLASS

In addition to good optical quality (undistorted transparency) and strength and weathering resistance, additional requirements are often placed on flat glass—requirements which cannot be met satisfactorily, if at all, by changing the glass composition. Architectural glazing exemplifies a case where it is necessary to have certain qualities such as heat reduction, light and sound protection, and increased resistance to breakage. A special safety requirement is the use of flat glass in vehicles. Finally, flat glass can also be given considerably

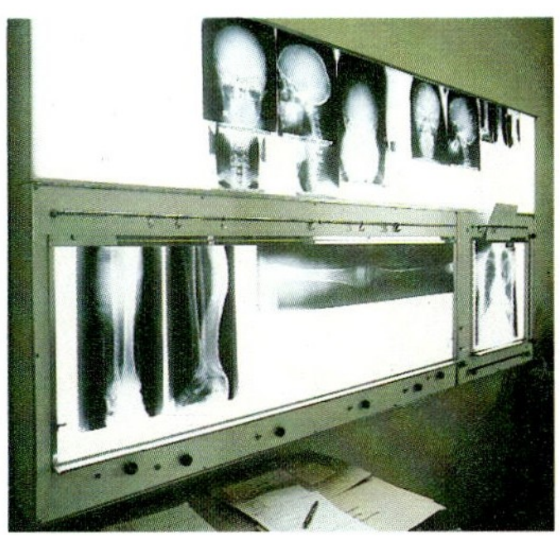

Fig. 4-10. Flashed opal glass for viewing X-ray negatives.

altered optical and decorative properties by special surface treatment. All of these things come under the category of industrial finishing of flat glass.

Glasses with Altered Radiation and Heat Transmission Characteristics (Sun Protection and Coldproofing)

Simple architectural glass transmits 75% to 90% of incident light depending on the thickness of the glass (see table on p. 00). Transmission decreases substantially at incident angles above 40° because of the increasing degree of reflection. The transfer of heat through glass always depends on the temperature difference between the inside and outside surfaces (thermal conductivity) and the thermal emission into the cooler of the two spaces. These two properties are identified by the thermal transfer coefficient, k. It shows the amount of heat per square meter per second which passes through the glass at a temperature difference of 1° C. It is measured in W/m^2K ($=0.86$ $kcal/m^2h°$). Together with the size and position of the window, it determines the energy consumption for heating and air conditioning of buildings. It also has a considerable influence on climatic comfort of interior spaces. Advantageous changes in the radiation transmission and k-value of flat glass can be achieved in different ways. There are three primary means of accomplishing this:

1. Multipane glazing
2. Glasses having increased absorption or dispersion of visible and infrared light
3. Flat glass with altered surface properties.

In many cases, products are made which combine these methods. (See Fig. 4-11.)

In Group 1, the insulating glasses have the most importance. They consist of two (sometimes three) panes which are either fused directly to one another at the edges or are joined by a metal frame in such a manner that the space between the glass (9-12 mm) is airtight. A desiccant is added to the space to prevent the inner surfaces of the glass from fogging as a result of temperature changes due to the

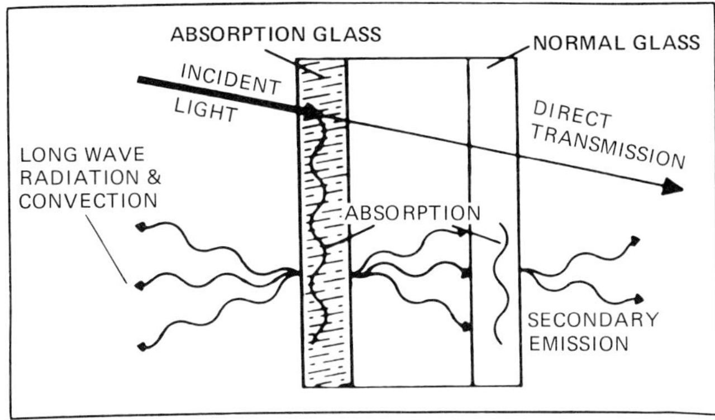

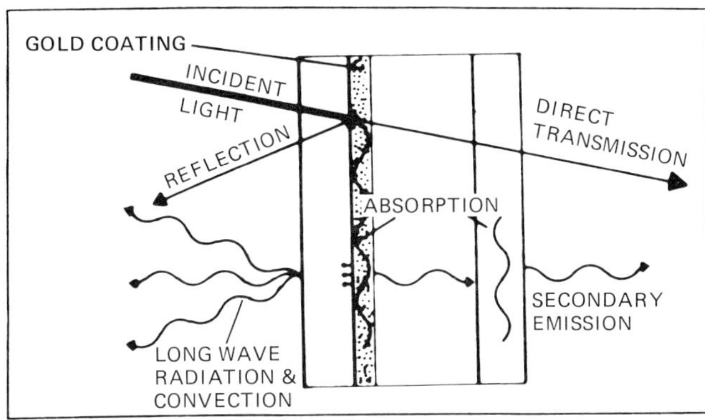

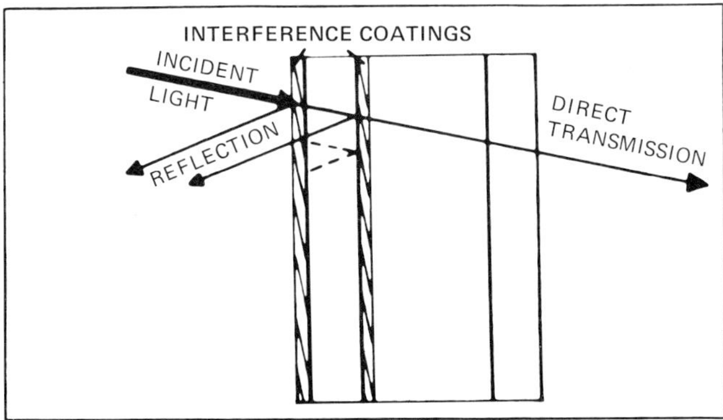

Fig. 4-11. Radiation distribution in solar protective double glazing with absorbing and reflecting panes.

weather. The thermal convection of the enclosed air—which is sometimes replaced by gases having a low thermal conductivity—is substantially reduced due to the narrow space. The k-value of a double
glazed insulating window having 6 mm thick panes and 12 mm air
gap is only 3.1 W/m^2K instead of 5.6 W/m^2K for single pane glazing.

Insulating glasses also serve to reduce street noise. The sound-
dampening effect of a window is measured in decibels. This unit
expresses the phon by which the sound level is reduced when traversing from the outer side to the inner side. A single pane window has a
sound dampening effect of about 25 decibels; that of an insulating
window achieves 30 to 40 decibels, however. Loud street noise (60 to
80 phon) is dampened by almost one-half. The effectiveness of insulating glass is also dependent on other factors, such as the angle of
incidence and the frequency (number of oscillations) of the sound.

Among the flat glasses with increased absorption listed in Group 2,
glasses having additions of bivalent iron oxide have become important because they are only slightly transparent in the near infrared.
Incident sunlight, and especially the heat in the 0.7-2.5 μm range, is
absorbed by it. Since absorption sharply increases above 0.6 μm, the
transmitted white light takes on a noticeably green tint, which can be
corrected to an extent by the addition of other absorbing components.
The k-value of the glass, on the other hand, is not noticeably influenced by the absorption properties. In sunlight, the glass warms to a
temperature at which the heat radiation into the environment is equal
to the energy gain of the absorbed radiation. Such glass gives limited
sun protection as building glazing. It is much more effective in vehicles due to the cooling effect of the wind. Absorbing glass must be
tempered when used in buildings (see Safety Glass, below), otherwise there is danger of breakage caused by localized stress from uneven heating of the window surface (if it stands partially in a shadow,
for instance). Similar problems arise with other tinted glasses used in
building glazings.

To dampen the brightness of large windows, gray and bronze colored tinted glasses have been developed. There are also special
yellow tinted glasses on the market which are opaque to UV and

shortwave blue light. These are used to inhibit the bleaching effect of sunlight on materials, paintings, etc.

Double panes between which a plastic film is sandwiched can also reduce transmission. This can be achieved by adding absorbing materials to the film. This is how windshields with gradually increasing glare-reducing tint towards the top are made. UV-radiation can also be better eliminated through the use of an imbedded UV-absorbing film than it can with tinted glass. The chief importance of these laminated glasses lies in their use as safety glass. (See Safety Glass, below.) High-dispersion, translucent dual-paned glasses are also manufactured with fiberglass filled spaces and sealed like insulating windows. Such a product is especially useful for thermal insulation.

Group 3 offers numerous possibilities for altering the radiation and thermal transmission of glass. By means of thin film technology for coating large surfaces, the reflection and absorption characteristics of glass surfaces can be altered in such a manner that energy conservation requirements and optimum air-conditioning requirements can be extensively achieved. For solar control, normal flat glass can be coated with metallic and/or oxide layers in such a manner that a high degree of light transmission can be achieved combined with a considerable amount of heat reflection (40%-60%). The effect is partially based on the high reflectivity of precious metals and partially on optical interference effects of highly refractive coatings, the thickness of which are usually about a quarter of the average wavelength of the radiation. The coatings are normally applied by vapor deposition, or sputtering under vacuum, or by dipping in or spraying with solutions, followed by firing. Due to their insufficient mechanical resistance, vacuum deposited precious metals require a protective coating with high abrasion resistance. Among some of the materials used to accomplish this are special borosilicate- and calcium-aluminosilicate glasses (by Schott) which provide highly durable protective coatings when vaporized by electronic bombardment in a vacuum. Methods of (nonelectrical) electroless metallization (see Reflective Flat Glasses, Chap. 5) are also used, particularly for the production of solar protective coatings. The usual configuration for solar protective double glazing combines a coated pane with an uncoated pane. The cooling energy in air-conditioned buildings is substantially reduced by this means. (See Fig. 4-12.)

To increase the capability of glass to insulate against cold temperatures, that is, to reduce the k-value of windows, thin film coatings are also used if they have a high degree of electron conductivity. Precious metals and several semiconducting oxides (tin oxide, indium oxide) can be used to accomplish this. In the case of the semiconducting oxides, this can be achieved by vacuum processes as well as by precipitation from vapor-forming or liquid compositions combined with concurrent or subsequent heating of the glass. The coatings allow more than 70% light transmission with a reflectivity in the infrared range of up to 90%, so that the thermal emissivity of the glass falls to between a fifth and a tenth of the normal value, thereby significantly reducing the k-value.

Instead of coating the glass itself, cost considerations have brought about the use of plastic films which adhere to the internal surface of the glass. Up to now, metallized films (with Al, Ag, or Au) have

Fig. 4-12. Hessische Landesbank in Frankfurt Main with *Itox* solar reflective glass.

been used almost exclusively. Their light transmission usually lies below 50%. Vacuum coating processes are mainly used for metallizing the films because they are economical and require no great amount of heat.

Antireflection Glasses

When observing items behind glass, like pictures, display items, measuring instruments, etc., a disturbing reflection of light often appears at the glass surface. About 8% of incident perpendicular light is reflected in such cases, and it increases as the angle of incidence becomes more oblique. This can be eliminated in several ways, for example, by fine etching (*silk matting*) of the surface which diffuses the reflected light, or by applying antireflective interference coatings having defined refractive indices and thicknesses. Interference coatings are applied by usual methods or by leaching techniques. Silk-matt etched glasses are suitable for pictures only if they are contact mounted. Otherwise, the contrast is impaired by the scattering effect which increases with the distance between the glass and the picture. It is widely used as *anti-Newton-ring glass* in slide projectors, whereby the matted surface inhibits the appearance of Newtonian interference colors upon contact with the slide.

Leaching can only be carried out successfully on certain special glasses and is therefore rarely used. An example of its use is in large antireflection blocks of radiation shielding glass. (See Special Optical Glasses, Chap. 6.) On the other hand, normal flat glass up to several square meters in area can be coated on both sides with a triple interference coating in a process developed by Schott. This reduces the visual reflection of light to about one-tenth of the normal value. Since the coating causes no noticeable absorption and dispersion losses, the reduction of reflection is combined with equivalent increased light transmission. This effect is important in optical systems such as camera lenses. Such antireflection coatings are almost always vacuum deposited. They are also used to reduce reflection and glare in spectacles.

Reflective Flat Glasses

A high-quality defect-free flat glass is necessary for making mirrors which offer the observers a natural, undistorted image. Float glass is the most economical glass to use to meet these requirements.

The Manufacture of Mirror Glass.   Washed glass plates are first treated on the side to be mirrored with a stannous chloride solution. This eases the silvering process by activating the glass surface. Then a silvering solution made of silver nitrate, ammonia, caustic soda or caustic potash, and distilled water together with a reducing agent made of dissolved glucose are sprayed onto the surface of the glass. The reducing agent causes the formation of silver nuclei which combine with the tin from the pretreatment and form a solid, fine crystalline silver film. This coating is not transparent at over 0.5 μm. The silver layer is usually allowed to grow to a minimum thickness of 0.01 mm. Then a similar process is used to coat the silver with a layer of copper. After drying, it receives two coats of protective lacquer. This is all accomplished in fully automated processes today. Formerly, the silvering solution was poured or rinsed over the glass.

The nonelectrical metallization of glass by wet processing is similarly applicable for gold and copper coatings. In strongly reducing solutions (preferably containing hypophosphites) it is even possible to deposit metals of the iron group (particularly nickel and nickel-copper alloys) as mirror surfaces. The attainable degree of reflection is generally less than 50% for such coatings. Similar processes are used in the United States for manufacturing solar protection glass. Very thin coatings having 15%-30% light transmission and 40%-25% reflection are used for this. The remainder of the radiation in such glass is absorbed by the coating.

The Uses of Mirrors.   Most mirror products are used in the home. Bathrooms and toilet areas are usually equipped with mirrors. Bedroom furniture usually includes mirrors. Decorative mirrors can also be made of rolled glass, both colored and tinted. *Antique mirrors* are so designated when the coating contains visible cracking so that it looks old. Mirrored walls in small rooms have an enlarging effect. We encounter rearview mirrors in automobiles daily and use them as safety related devices. By applying several oxide coatings using the dip coating process, interference semi-mirrors are made which result in antiglare rearview mirrors and *one-way mirrors* for observing brightly lit rooms from darkened areas. Self-service stores are often supervised this way. (See Fig. 4-13.)

Cold mirrors are manufactured by applying at least 12 optical interference coatings having alternating high and low refractive indices onto a heat-resisting glass. They are so constructed that only visible light is reflected to any great extent; thermal radiation is transmitted or absorbed. Chemical reducing processes are suitable for one-side mirror coating only because the layers are insufficiently durable without a good protective coating. Optical mirrors are either front-surface-coated with stable vapor deposits to prevent double reflection; or if rear-surface-coated, the front surfaces are antireflection coated.

Other Techniques of Finishing the Surface of Flat Glass

Many surface treatment procedures for flat glass depend on fashion trends. For a long time, frosted designs were popular. To achieve this, hydrofluoric acid is poured onto the surface and then quickly rinsed off. Since the acid does not evenly attack the glass, designs resembling ice-ferns appear. A similar design results when glue is applied to a matt surfaced glass. When removed, the drying, shrinking glue tears off thin layers of glass, leaving an arctic design.

Fig. 4-13. Antiglare (and heatable) rearview mirror.

By etching with hydrofluoric acid, different degrees of frosting or matting are achieved. Frosted glass is used in interior fittings and in windows where transparency is not desirable.

Glass surfaces can be made rough or matt by sandblasting. By covering the surface beforehand with stencil masks, the glass surface can reproduce representational or abstract decoration.

A further development is the engraving of flat glass using a flexible shaft. This entails a small grinding wheel driven by a flexible shaft to transmit desired contours, freehand, onto the glass surface.

Safety Glass

As daily experience has shown us, flat glass often breaks under a slight mechanical load such as impact or pressure as well as with quickly changing temperatures. The sharp fragments of broken glass can cause serious injury. Special treatment can render the glass less susceptible to breaking and considerably reduce the chance of injury. The result of such processing is *safety glass*.

Tempered (or toughened) Glass.   During production, cut pieces of flat glass are suspended above or placed horizontally into processing equipment where they are quickly heated to about 150° C above the transformation temperature. (See Technical Identification, Chap. 4.) Immediately afterwards, the glass is blasted with cold air in an appropriately designed system of jets. As a result of this quenching process, the internal glass mass cools more slowly than the external surface, and it is still contracting after the surface has already solidified. Since they are bound together, the outer surface layer is subject to compression and the inside to tension. The procedure itself is called thermal or physical toughening of glass; the finished product is called *tempered glass* or *monolithic safety glass*. The most important factors affecting the degree of prestressing are the thermal expansion of the glass above and below the transformation temperature (T_g), its modulus of elasticity, and the temperature difference above T_g which exists between the outer surface and the internal mass of the glass. The compressive stress obtained is approximately three times that of the normal bending strength, and it increases with glass thickness. A bending load has therefore that much less effectiveness, resulting in increased bending strength. When thermally toughened safety glass is impacted, e.g., when a windshield is hit by a stone while driving, or if a car is involved in an accident, it breaks into many small, almost regularly shaped pieces of glass having no sharp edges. The size of the particles of glass can be controlled in advance by modifying the toughening procedure. This is important for automotive windshields in order to limit the impedance to the driver's vision. Often monolithic safety glass for windshields is equipped with so-called vision islands which crack into larger pieces than the remainder of the glass. They provide reasonably good vision until the damaged windshield can be replaced. There are several German Industrial Standards (German Standards Institute) test procedures for monolithic safety glass. (See Fig. 4-14.)

All required flat glasses can be prestressed. Usually today float glass is toughened for use in vehicles. Monolithic safety glass, along with other types of safety glass, is usually produced by the large flat glass companies which manufacture the raw glass itself. During the same process, the tempered glass is also bent since front, rear, and side windows are often curved. Bending must precede tempering because the tempered condition would disappear at the bending temper-

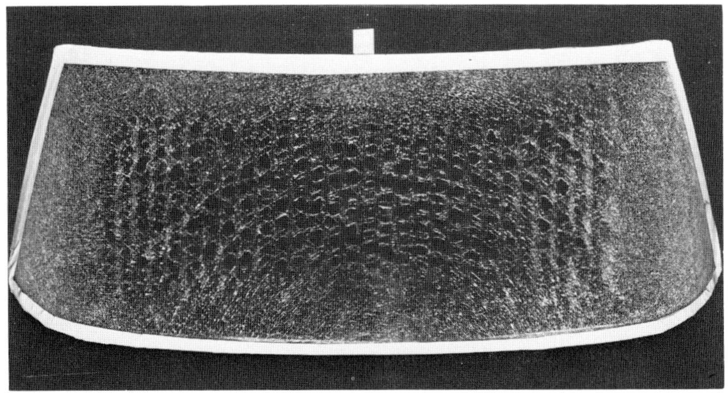

Fig. 4-14. Breaking pattern of a windshield made of tempered safety glass.

ature. In office buildings, tempered flat glass is used in the form of glass doors, room dividers, elevator glazing, or stairway landings. Ornamental rolled glass is also popular for such use if clear vision is not desired. (The possibility of processing safety glass into multipane glazing has already been discussed in Processed Flat Glass, above.)

To clad the facades of large buildings, the wall surfaces are often covered with tempered flat glass for decorative purposes. The rear surface of such glass is often coated with a color-coordinated layer of enamel. To achieve this, the glass is dusted with enamel frit before entering the toughening furnace. The enamel melts when heated and binds itself to the glass surface without influencing the prestressing process. In doing this, not only does a decorative effect result, but greater resistance to weathering is achieved.

Thermally toughened glasses cannot be further processed. Therefore, they must be cut to final size before tempering. The same holds true for drilling and edge work. This is especially important before fixing handles and locks to glass doors.

Chemical Toughening. The chemical prestressing of glass is characterized by compressive stressing of the glass surface by modifying the surface composition without changing the inner body of the glass. In 1891, Otto Schott succeeded in producing a very strong glass by coating a glass with a low coefficient of thermal expansion over a glass with a high expansion. This process is related to those used today in the production of tableware and flashed opal glass. Methods of chemical prestressing are based on ion exchange. The ions of the individual chemical elements have different radii and are arranged at different distances from one another. For example, if a sodium-containing glass is slowly heated to just below the transformation temperature in a molten potassium salt, the sodium ions migrate from the glass into the salt melt, and the potassium ions leave the salt melt and enter into the glass surface. This is called *ion exchange*. Since the potassium ions which penetrate the glass have a 30% larger radius, a shortage of space arises in the glass surface, which results in a compressive surface layer. The exchange zone must be at least 0.1 mm thick in order to increase the strength by 5-6 times. Glasses toughened by ion exchange are used for special purposes in the aircraft industry, for centrifuge tubes, data processing, and the lighting industry. The same process is used for toughening ophthalmic lenses.

Laminated Glass.   Laminated (or *compound*) safety glass consists of two or more panes (usually float glass) which are joined with a visco-elastic layer of plastic. The permanent joining of the glasses occurs in a pressurized vessel called an autoclave where simultaneous heating of the preprocessed "sandwich" takes place.

When laminated safety glass is broken, the broken pieces of glass remain bound to the internal tear-resistant plastic layer. The pieces are not freed, and the broken sheet remains transparent.

As opposed to prestressed safety glass, laminated glass can be further processed. It can be cut, drilled and edge-worked.

Laminated safety glass is used in windows of buildings when they are required to be enter- or escape- or explosion-proof, in spandrels, in walls, room dividers, and roofs. In many countries, auto windshields must by law be made of laminated glass. Legislation not withstanding, it is widely used anyway. It is used as a protective glass in machinery, instruments, and television sets. In addition, there are several special forms of laminated glass which should be mentioned.

Armor Plate Glass   consists of at least four layers and is at least 25 mm thick. Such glasses are bullet-resistant to handguns. Armor plate glass over 60 mm thick is bulletproof. Cashier's offices, bank counters, jewelry store windows, or special transport vehicles can be equipped with armor plate glass. (See Fig. 4-15.)

Wired Compound Glass   describes the glass which results when steel filaments are imbedded between the sheets of glass to increase security.

Alarm Glass   is, to an extent, a further development of wired compound glass. Imbedded in the intermediate layer are 0.1 mm thick wires which form an electrical circuit. When this glass is broken, the circuit is broken, thereby setting off an alarm at the same time.

Fig. 4-15. Testing bullet-resisting glass.

Heated Glass   is always useful when moisture and ice formation must be avoided. This can be the case in windshields and back windows in cars, in airport control towers, in flower shops, in cold storage houses, or in indoor swimming pools. This effect is achieved by one of two ways. Either the intermediate layer contains barely visible wires which warm when connected to an electrical source, or the glass is coated with a transparent electrically conducting surface layer which warms similarly.

Antenna Glass   is also made by two methods. In one method a thin wire of a certain length and shape is imbedded in the intermediate layer of a windshield. In the other method, the antenna function is carried out by a conductive thread of silver which is fused into the inside surface of a windshield.

Colored Laminated Glass   (both fully and partially colored) is used in buildings and vehicles. The films between the sheets of glass are colored. They can have heat reducing, UV-absorbing, reflecting, and anti-glare effects. (See Processed Flat Glass, above.)

Laminated Glass Used in Combination with Other Glass. The risk of injury becomes smaller when the inside sheet of a laminated glass is thinner (1.5 mm for example). The same purpose can be accomplished by using a tempered glass on the inside. If safety is to be combined with heat and sound insulation, laminated glass is used in combination with normal glass or another laminated glass to make insulating glass. Very thin laminated glass is also used in safety spectacles and face masks in helmets because they are less susceptible to scratching than plastics.

Since widely differing glass types can be combined in multipane glazing, there are many practical applications where several types of processed flat glass are used together. Insulating safety glass is found, for example, in railroad cars, sometimes in conjunction with sun protection glass. Gatekeeper's houses can be equipped with multipane glazing containing safety glass or even bullet-resistant glass. Control towers of airports are glazed with large areas of glass to give maximum vision. Air traffic controllers are assured of bearable climatic conditions through the use of a sun protection glass as a component in multipane glazing.

Fire-resistant Glass

The increased use of glass in buildings (in facades, walls, and partitions) brings with it an increased fire risk. Normal flat glasses shatter after only a short time when subjected to one-sided heat, large pieces of glass fall out of windows, and the threat of the quick spread of fire to another story increases. Formerly, there was only one method of retarding the shattering of glass during a fire, namely, a wire inlay inside 6-8 mm thick panes which prevents the glass from falling from a window when it breaks.

Like other construction materials, fire-resistant glazings are also classified into fire-resistance classes G and F in the internationally specified fire test. According to the test described in DIN 4102, Part 5, glazing which resists the passage of flames and smoke for 30, 60, 90, 120, or 180 minutes are categorized as G 30, G 60, G 90, G 120, or G 180. Analogously, classes F 30, F 60, F 90, F 120, and F 180 represent those glazings which impede radiant heat for at least 30, 60, 90, 120, and 180 minutes, respectively, and which do not heat up on the unexposed side to more than an average of 140° C above the initial temperature. Additionally, such glazings must withstand certain steel ball impact tests. (See Fig. 4-16.)

According to this classification system, the above-mentioned wired glass with a concrete frame and with under 2 square meters area is in class G 60. For visual reasons, applications are mainly limited to such things as partition walls, doors, and skylights.

New developments in fire-resistant glass have resulted in wire-free glasses suitable for use in windows. *Pyran* (Schott) is a thermally prestressed borosilicate glass which is officially approved for class G 90 in 2 × 1 square meter sizes and class G 120 in 1 × 1 square meter sizes. Because it is prestressed, it breaks into relatively small pieces upon impact. Several flat glass manufacturers have also introduced double glazings into the market, the inner space of which is filled with a material which turns to foam in case of fire thereby acting as a heat shield. They are in classes F 30 to F 90.

Fig. 4-16. Fire-resistant glass under test.

5. Hollowware and Glass Tubing

The most widely distributed glass products belong to the hollowware family. In the Federal Republic of Germany, 75% of the glass produced and 50% of the revenue from glass is accounted for by hollowware. Hollowware is encountered everywhere. In the broadest sense, these products are consumer goods when speaking of such things as bottles, or consumer durables when speaking of drinking glasses or glass lamps.

Most hollowware is made of soda-lime glass. Crystal and lead crystal and numerous other special purpose glasses are exceptions.

The processing and finishing of hollowware are important fields in their own right and result in a wide variety of products. Their starting material is hollowware manufactured by the glass factory.

THE MOST IMPORTANT TYPES OF HOLLOWWARE

The huge hollowware sector can be divided into several groups, according to different criteria. One of these is the manufacturing process, which can be subdivided into mouth blowing, machine blowing, and pressing to name the most important.

Another classification criterion is the chemico-physical properties. This results in such classifications as *bottle glass, semiwhite hollowware, crystal glass, lead crystal,* and so on.

The most common method of classification is according to use. This is how worldwide statistics on the industry are kept. This classification includes *container glass* (bottles, jars, medical, and packaging glass), *tableware* (drinking glasses, and other glassware for the table, the kitchen, and the home), and *construction hollowware* (glass building blocks, etc.). *Medico-technical hollowware* and *illuminating glasses* fall primarily within the realm of the special glasses (Chapter 6) and are discussed there.

Each of these types of hollowware will be discussed individually. However, since many of them use the same manufacturing processes, the following section will be devoted to this subject.

THE SHAPING OF HOLLOWWARE

The first known hollow glass body was fashioned in ancient Egypt by coating a shaped sand core with viscous glass. (See Fig. 1-1.)

The real beginning of the manufacture of hollowware did not start however until the invention of the glassblower's pipe in about 200 B.C. The technique of shaping glass with a glass blowpipe has remained in the originally developed form to the present day. Even the functional methods of modern blowing machinery are based on the manual techniques, though they may be hard to recognize.

The Mouth Blowing Process

The glassworker's blowpipe is a steel tube with a wooden handle and mouthpiece at one end. It is about 1.5 meters long. At the other end is the *nose* (or gathering head), an extension which picks up a *gather* when dipped into molten glass. By rotating and swinging the pipe, the glass is prevented from running off while it cools. The gather can be dipped into the melt again to get more glass, depending upon how much is needed. (See Fig. 5-1.)

The first hollow body, the *parison*, appears when giving a short puff into the pipe. The external shape can be changed by turning in a hollowed, water-soaked beech wood block (*blocking*) or on an iron plate, whereby the gather becomes more viscous as it cools.

The subsequent reheating in the furnace, along with rotating and swinging of the pipe allows the temperature difference to equalize before the article takes its final shape. (See Fig. 5-2.)

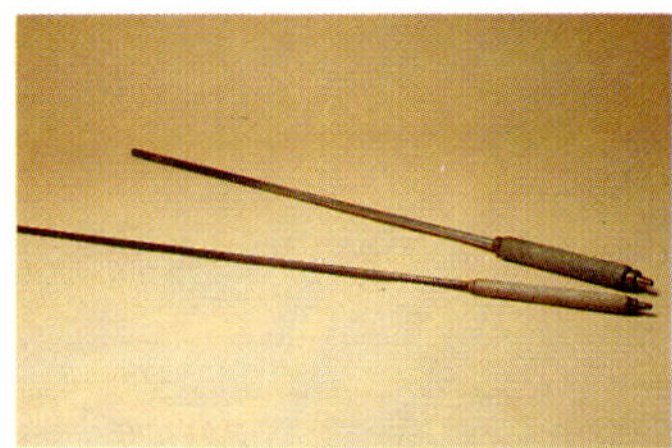

Fig. 5-1. Glassblowing pipes.

Fig. 5-2. Glassmaker at work.

The final blowing of individually shaped articles is completely manual, and only a few special pieces of equipment are used, such as a wooden board (*paddle*) and tongs. However, the glass is more often blown into a mould which allows the production of many similar hollow items.

Originally made from water-soaked beech wood, moulds are also made of graphite or cast iron for large volume production. A coating mixture of sawdust and a binding agent are "pasted" to the inside surfaces of cast-iron molds. This paste is burnt-in and soaked in water before each use. The life of the paste is about 500 operations, after which it must be renewed. The life of a metal mold is practically unlimited.

On contact with the rotating glass, a steam cushion forms in the water-soaked mold; and due to retarded cooling, an even thin-walled hollow body can be blown. This results in a brilliant surface like that obtained in free blowing, which could otherwise not be achieved using other mold forming techniques. (See Fig. 5-3.)

Fig. 5-3. Free-form blowing.

Other processes which must be carried out before hollowware is ready for use include the attaching and shaping of another glass-gather for a handle, stem, or base; the separation from the pipe, transfer into a lehr; and cutting off the excess glass. (See Fig. 5-4.)

The various steps in the manual production of hollowware are performed by specialist craftsmen, on whom the highest demands of physical condition, manual dexterity, and artistic talent are placed. Thus one can more easily understand that only high quality tableware (crystal goblets, for example) and low quantity technical articles having difficult shapes are handmade today by an ever-decreasing number of glassmakers. The low costs of a multitude of hollowware products which we encounter every day (beverage bottles and light bulbs, for example) are only made possible by large-scale automated production.

Machine Blowing

The development of mechanical production processes led to the first automated blowing machines, made by Michael Owens in 1903, and development continues to this day.

The first machine-made products were bottles (packaging glass). Due to the lower surface quality requirements, paste molds were eliminated. This made fast heat dissipation possible, thereby increasing machine output. The hourly output increased from about 17.5 to 90 pieces per finishing mold, and it has risen to about 900 and more in the most modern machines. Several years later it became possible to automate the rotating mold (paste mold) process.

Fig. 5-4. Attaching a handle.

All automated blowing processes follow the manual blowing format after formation of the gather: 1) preforming (forming the parison), 2) reheating (equalizing the temperature differences), 3) final blowing. Although there are many different machines today, they can be classified into three basic types:

1. Continuous rotary machines, used partly in manufacturing container glass, but primarily in paste-mold processes.
2. Adjacent, complete sections, each independently driven, like the I-S machine (Independent Section Machine), which dominates the container glass industry. (See Fig. 5-5.)
3. The chain or ribbon-type machines, which are used to make light bulbs in the paste-mold process.

Following are descriptions of the most important blowing processes. Several of them have been technologically outdated, but are still used in small market segments.

Suck-Blow Process.   The Owens bottle machine belongs in this category. The glass gather is sucked by the parison (preform) mold from the melt surface and is cut with automatic shears. A short plunger or *core* projecting into the mold produces a hollow space in the glass which is enlarged by blowing when the parison mold is

Fig. 5-5. I-S machine for container production.

opened during reheating. For final blowing, the parison which is held by the neck mold, swings into the blow (finishing) mold. (See Fig. 5-6.) This production method was replaced by the blow-blow process with the advent of the gob feeder.

Blow-Blow Process.   The elongated gob which is formed by a gob feeder falls into the parison mold and is blown from below against the bottom of the inverted parison mold. The rest of the process is like the suck-blow process. (See Fig. 5-7.)

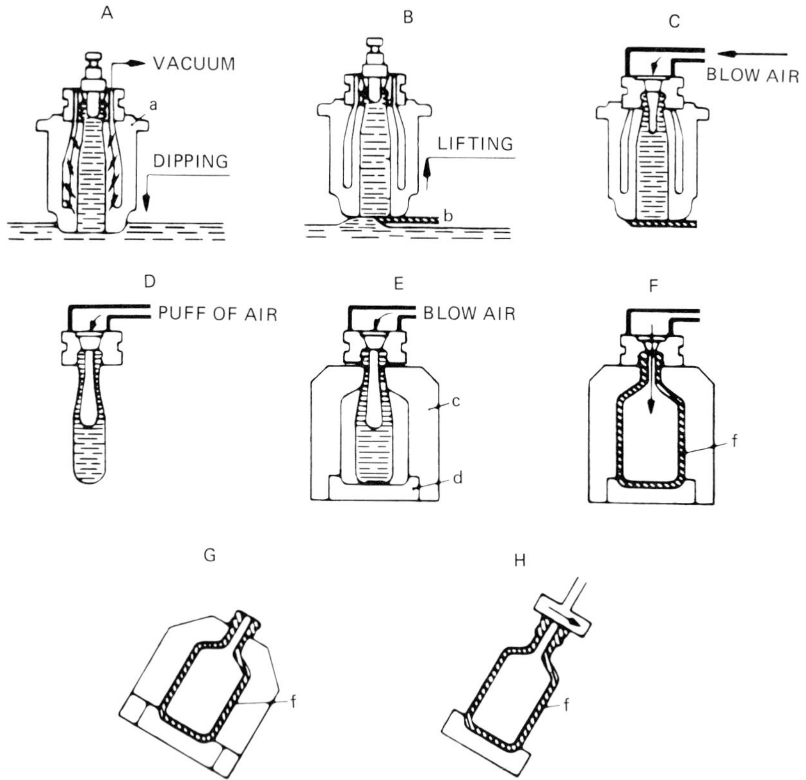

Fig. 5-6. Manufacturing hollowware in a suck-blow machine. a) parison mold, b) shears, c) finishing mold, d) bottom mold, f) finished bottle. A) gathering the glass by dipping in the glass bath and suction, B) Cutting off the glass, C) Attachment to air supply, D) Initial blow, reheating, and lengthening, E) placement in the finishing mold, F) final blow, G) discharging the finishing mold and cooling, H) removal of the finished bottle.

Fig. 5-7. Container glass produced by the blow-blow process.

In addition to blow-blow, the most modern development is the press-blow process. In some cases both processes can be used on the same machines with some refitting.

Press-Blow Process.   At the preform station, the parison is not blown, but pressed by a plunger extending from the neck mold. This process permits high production rates because the parison is more quickly cooled; and because it gives more controlled glass distribution, it allows the manufacture of thin-walled glass containers, such as disposable bottles.

In the manufacture of very narrow mouth diameters, not all mechanical and thermal problems have been mastered. Both of the preceding processes yield a surface quality and wall thickness that are not acceptable for a number of glass products. These products are reserved for the automatic paste-mold process.

Rotary-Mold (Paste Mold) Process.   The formation of the gather can be accomplished by suction or gob feeding. The lens-shaped or hollow-bodied gather, which is preformed by suction or by pressing of a gob, is held by a metal ring on which a blowhead is located. Several puffs of air, combined with gravity, cause the parison to lengthen until the water-saturated paste mold closes; and the final blow begins with simultaneous counter rotation of the glass and the mold. The solidified glass which comes in contact with the retaining ring and the blowhead is not usable and is separated like the cap produced in manual production.

In addition to the blowing processes, several other important methods for manufacturing hollowware are in use.

Pressing

Unlike the blowing processes, in pressing, the glass-gather touches all parts of the metallic mold material. As a rule, the pressing mold consists of three parts: the (hollow) mold; a plunger, which fits into the mold leaving a space representing the thickness of the glass wall; and a sealing ring which guides the plunger when being removed from the mold.

A gob is fed into the mold and is hydraulically or pneumatically pressed by the ring-guided plunger until the glass is pressed into all areas of the mold. After solidification, the plunger is withdrawn. Most pressing machines operate on turntables which have 4-20 or more molds. The turntables take the glass step-by-step through the loading, pressing, cooling, and other processing stations to the removal point.

Typical pressed items are heat-resistant household glassware, drinking glasses, lamp globes, and parts for television tubes. (See Fig. 5-8.)

Spinning (Centrifuging)

This shaping procedure is fairly new, and not well-known. It is based on the principle that a rotating liquid will take up a surface shape of a rotation paraboloid thus forming a hollow space.

Fig. 5-8. Automated pressing of dishes.

In its application, an extremely liquid glass-gather is fed into a steel mold which is then rotated at the required rate. At high revolution rates, the parabola becomes very steep and is almost cylindrical. When the glass has cooled sufficiently to harden, the rotation is stopped, and the glass is removed. An example of a spun article is the borosilicate glass column section used in chemical plants. They can be made with diameters of up to one meter. Other axially symmetrical items such as funnels for television tubes are sometimes also made by spinning.

THE DRAWING PROCESS FOR GLASS TUBING

Glass tubing is not exactly a hollowware product, but it merits mention at this point, because a number of tubing products are made into containers, such as ampoules, vials, fluorescent light tubes, etc. It is rarely manually produced anymore.

The most widely used method of continuous production of tubing is named after the inventor Danner, an American engineer who invented the process in 1912 (US Patent 1218598 of Libbey Glass Co.). A continuous strand of molten glass flows onto a slightly angled, slowly rotating fire clay core, called the *Danner mandrel*. At the lower end of the mandrel, a hollow bulb forms from which the tubing is drawn. Air is introduced through the hollow mandrel shaft maintaining a hollow space in the glass. After being redirected horizontally, the hardening tube is transported on a roller track to the drawing machine, behind which it is cut into 1.5 meter lengths. The output of such a machine can be 3 meters per second or more. (See Fig. 5-9.)

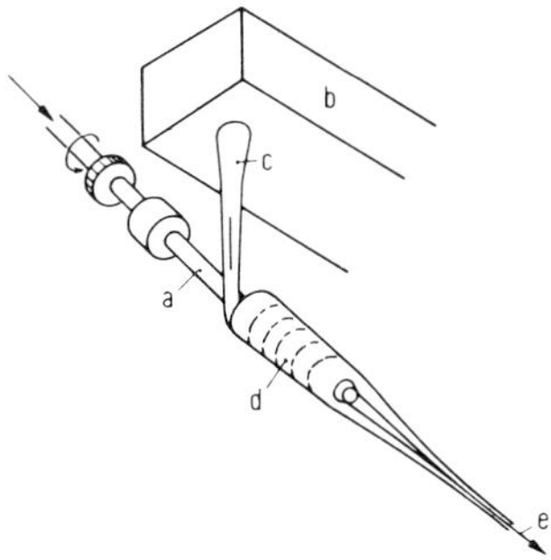

Fig. 5-9. Danner tube drawing process: a) rotating mandrel, b) furnace forehearth, c) glass flowing from exit ring, d) wound glass mass, e) draw direction.

Other Tube Drawing Processes

The *Vello process* is as important as the Danner process and has about the same output. The glass from the furnace forehearth flows downward through an orifice (the *ring*) and the hollow space in the glass is maintained by a pipe with conical opening (the *bell*) located within the ring. The tube, which is still soft, is redirected horizontally and is drawn off along a roller track, cooled, and cut as in the Danner process.

In the *Schuller-Metz (Updraw) process*, the glass tube is drawn vertically upward from a rotating bowl. The drawing area is shielded by a rotating ceramic cylinder one end of which is submerged in the glass mass. The formation of the hollow space is by means of an air jet which is placed below the surface of the glass. Tubing with thick walls and considerable diameters is made by this process.

In a variation of the Vello method, the glass is drawn downward (*Downdraw process*) through a vacuum chamber. The glass passes through an iris diaphragm (a closure which can be adjusted for different sized circular openings) which maintains the seal. At a drawing speed of several meters per second, tubing with a diameter of up to 360 mm can be produced by this method.

If tubing having close tolerances is needed for special purposes, such as in bottling filling machines and flowmeters, this can be accomplished by shrinking the heated tubing under partial vacuum on a steel mandrel having the desired diameter (the Schott "KPG" process).

According to use or further processing requirements, several different glass types are used for producing tubing. Soda-lime glass is suitable for fluorescent light tubing or small lamps; electronic tubes require highly insulating lead glass; chemical resistant glass is required for pharmaceutical uses; and borosilicate glass is used in the chemical industry.

FINISHING OF HOLLOWWARE

The finishing of hollowware includes all stages necessary to produce a new item from a preshaped glass. As a rule, this occurs by reshaping under the influence of heat. All kinds of glass tubing are used to make goods in the hollowware finishing industry.

Lamp Blowing (Lampworking)

This is what the manual process for finishing hollowware is called. The *lamp* is a gas torch which together with the Bunsen burner is used to heat the raw glass. The glass-blower (who should not be confused with the glassmaker who works the raw, molten glass from the melting furnace) needs temperatures between 600°-1700° C, depending upon the glass type. The tools he uses depend on the product to be made and the type of glass to be worked; the glass can be soft or hard, that is, melt easily or with difficulty. The glass-blower's chief aids are the glass cutting knife, the flaring tool, the carriage, the nippers, the forceps, and measuring tools. (See Fig. 5-10.)

Glass instruments and apparatus are typical examples of lampblown items. This includes such medical and veterinary equipment as syringes, blood sampling and testing devices, pipettes, and much more. In addition, apparatus for measuring pressure and flow of gases and liquids, for viscometry, electrochemistry, and water distillation can be mentioned. (See Fig. 5-11.) In food processing plants and dairies, there are glass instruments for a myriad of uses like acidity testers and volumetric equipment. In laboratories, we find condensers, retorts, burettes, drop counters, separators, and analysis glass. One important group in its own right is thermometry: thermometers

Fig. 5-10. Glassblower with *lamp.*

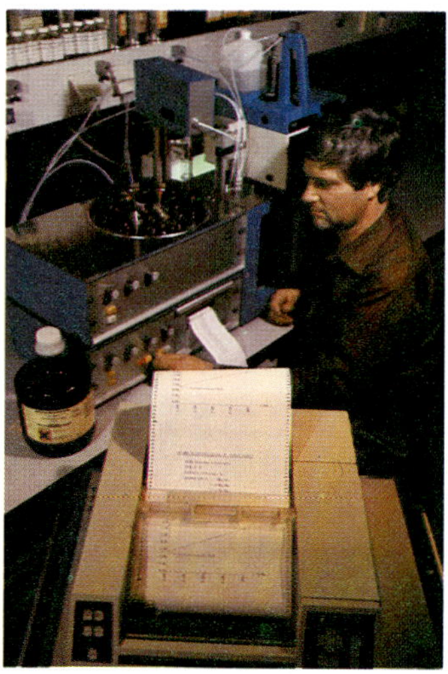

Fig. 5-11. Computer controlled diluting viscometer.

for health care, for industry, for the home, and for science. Lamp blowing is also fast becoming a process for artistic creation. (See Fig. 5-12.)

Industrial Hollowware Processing

For packaging its products, the pharmaceutical industry needs huge quantities of ampoules, vials, small bottles and tablet container tubes

Fig. 5-12. Lampblown art glass by Kurt Wallstab.

made from glass tubing. In order to be compatible with the high-output filling machines in the pharmaceutical industry, they must be made to precise measurements in automatic equipment. Simple glass apparatus can now also be made in semi-automated processes, especially where the apparatus has been standardized. Glass funnels are a good example. (See Fig. 5-13.)

Insulating Vessels

Veiled behind this term, normally encountered only in industry and trade, is a useful consumer item available in nearly every household—the thermos flask. It operates on the physical principle of heat and cold insulation by a double-walled glass container. It consists of two glass vessels which are fused in such a way that a hollow space remains between them. In another process, double-walled cylinders are blown at the furnace. Both glass bodies are later silvered on their interspace surfaces through a glass tube attached at the vessel base. (See Reflective Flat Glasses, Chap. 4.) Finally, the air is pumped out, and the tube is sealed off. The hollow space inhibits the heat exchange between the contents of the container and the environment, and the mirrored surfaces additionally help to prevent thermal radiation. The insulating vessel is normally placed inside a protective container made of plastic, sheet iron, aluminum, stainless steel, or brass.

Fig. 5-13. Production of small pharmaceutical bottles.

Many insulating vessels are also decorated. According to use, insulating vessels are referred to as insulating flasks, insulating jugs, food containers (with wide openings), and ice containers. So-called Dewar flasks are large capacity insulating vessels used for the storage of liquid gases and other laboratory uses.

Glass Jewelry

Glass beads are made by cutting off small pieces of colored tubing and rotating them in a heated drum. Charcoal, graphite or gypsum are added to prevent the pieces from fusing together. Pearl reproductions are made by blowing small, hollow spheres from thin glass tubing and filling them with a mother-of-pearl colored paste. Glass pearls are strung and made into necklaces, bracelets, and rosaries. Small, colorful, glittering pearls, called *Rocailles* are used for beading in clothing. Artificial gemstones are also pressed and ground from colored glass rods. By coating them with thin, transparent layers, polished glass pearls can be made iridescent. (See Iridescent Glass, Chap. 5.) Glass artisans from Gablonz in northern Bohemia are world-famous for their glass jewelry. For this reason it is sometimes referred to as *Gablonzware*. New forms of fashion jewelry made from it are constantly appearing on the market.

CONTAINER GLASS

This is a general term denoting all hollowware used for packaging, storage, preserving and transportation of beverages and other liquids, foods, chemicals, pharmaceuticals, and cosmetics. Container glass is always manufactured in the glassworks. That is, it is always made from molten glass. In this way it is distinguishable from vials and ampoules which are manufactured in the hollowware finishing sector. (See Fig. 5-14.)

The importance of glass as a packaging material lies in its ability to be readily shaped. It can be shaped in such a way that a firm or a product can be identified by the container alone. Take, for example, wines and spirits. The consumer can immediately recognize the vine-

Fig. 5-14. Typical glass containers.

yard district of the wine from the shape of the bottle. Similarly, different types of spirits can be recognized through the shape of the bottle in which they are packaged.

The properties of glass are even more important reasons for its use as a packaging material. Glass is odorless, impermeable, physically and chemically stable, sufficiently acid-resistant (even extremely so—Special Glasses, Chap. 6—except against hydrofluoric acid), resistant to alkalis, transparent, easy to clean, and hygienic. A disadvantage is its relative heaviness. But new developments have lessened this disadvantage. Glass containers have become increasingly lighter since it has become possible to manufacture them with thinner walls without affecting their mechanical stability.

Most glass containers are made from soda-lime glass. Relatively pure raw materials are used to melt colorless glass from which bottling and preserving jars, and bottles for beverages, cosmetics, cleaning materials, and noncritical pharmaceutical products are made. When there are special requirements for chemical resistance, however, special glasses must be used. (See Pharmaceutical Glass, Chap. 6.)

Green and brown are the most prevalent colored glass containers. Green glass, which transmits UV-light, is obtained by adding chromium oxide to the batch. Brown glass, which absorbs almost all UV-light, is obtained by adding iron sulfide, either directly as pyrites or by simultaneous addition of Na_2SO_4 and carbon (as a reducing agent which in the presence of iron produces so-called carbon amber. (For other colors, see Raw Materials, Chap. 2.) The color can serve as both a means of identification (e.g., in some regions brown glass signifies red wine) and protection from light, or it can have purely aesthetic reasons. The amount of coloring agents in a glass batch is usually less than 0.5%; and practically nothing is leached out by the contents of the bottle, even during long periods of storage.

Beverage Bottles

These are made by both blow-blow and press-blow processes. In West Germany, the lawmakers have legislated standard bottle sizes for various beverages. This protects the consumer from being deceived. For example, wine and liquor bottles are *full bottles* when their contents are 0.7 liters. A *half bottle*, on the other hand, is 0.35 liters. Beer is sold in 0.5 and 0.33 liter bottles. Other bottled products have other standard volumes. Bottles having contents of 1, 1.5, 2, 3, and additional full liter amounts are allowed for all products.

Legislators have determined in which cases a bottle may serve as a unit of measure. The precision of bottle dimensions is very important here, and the standards bureaus closely control this by taking random samples in the glass factory. Bottles marked with an *M* on the bottom are legally certified units of measure for the purpose of this legislation. Also stamped on the bottom is an official manufacturer's trademark and a rated volume. Thus beverage containers, unlike bar glasses for beer, spirits, etc., do not need a calibration mark.

Glass balloons or carboys can contain upwards of 60 liters. They are usually wrapped in wicker or straw and placed in metallic or wooden baskets (which is why they are sometimes referred to as *basket bottles*). Acids, alkalis, and other liquid chemicals are stored and transported in them.

In competing with lighter container materials, such as tinplate or plastic, manufacturers have long desired to reduce the weight of glass containers by making the walls thinner without affecting their mechanical strength. This has been achieved on the one hand by more uniform and reduced wall thickness, and on the other by reducing friction and minute surface fissures by hot vapor depositing of tin oxide in combination with a plastic coating on the outside surface. This has increased the bursting pressure by 2-3 times.

From a business standpoint, the one-way bottle was promoted as the best alternative to returnable bottles. The increased level of environmental awareness led to the recycling system, established by glass companies in 1975. Discarded bottles are added as cullet to the batch and remelted. This results in considerable savings in raw materials and energy. One difficulty associated with this program is the necessity to sort the discarded bottles according to color, since both clear and brown glass colorings are sensitive to impurities. Green glass, on the other hand, can accept other glass types without noticeable influence on the color.

Bottling Jars

Industrial and household preserving vessels (bottling jars) are usually made of clear glass. Clear vision of the contents of a jar is a good sales aid, especially for foods.

The chemical resistance of such glass is very important. The resistance to aqueous solutions corresponds to hydrolytic class 3 in DIN 12 111, which means it is high enough to withstand repeated boiling with its contents and that washing processes result in no noticeable changes in the surface of the glass such as clouding or staining. Due to the relatively high thermal expansion of soda-lime glass ($\alpha \approx 9 \times 10^{-6}$/K), rapid temperature changes must be avoided, especially in jars with thick walls or bases, because this can cause breakage.

GLASS TABLEWARE

This general and often confusing term comprises a group of hollow-ware products which are encountered in daily life and which have high demands placed upon them, especially from the design point of view. This glass is often referred to as household glassware.

The main members of the tableware family of products are drinking glasses, which comprise in terms of value about 60% of the tableware produced. The remaining 40% is accounted for by other glass accessories for the table and by articles which are used in the kitchen, the dwelling, and in offices.

Drinking glass sets include all glasses used in the household or commercially for individual drinks. Since these often have stems, they are usually called *stemware*. The term "set" denotes that they have basic designs (shapes or method of decoration) common to all glasses belonging to that set. (See Fig. 5-15.)

Fig. 5-15. The *Europa* stemware set by Schott-Zwiesel.

Typical tableware products, sometimes also called *giftware*, are ashtrays and smoking sets, table and floor vases, flower bowls, candle holders, and large glass dishes. Decorative glass items without specific functions, figurines, glass animals, and wall decorations made of glass also belong to this broad group. Simple drinking glasses for catering and daily domestic use are almost always machine made. It is also possible to produce stemware by machine, including crystalware and lead crystalware, which is only distinguishable from the handblown product by the expert. Such products do not pretend to replace manual production, which for complicated and individual designs is second to none. Instead, they ensure that a wide segment of the buying public can have high quality, mass-produced products available to it at a reasonable price.

Breakdown of Tableware by Glass Type

Compositionally, there are basically three main groups of tableware. By far the largest group is soda-lime glass. The main producers all have slight variations of this glass. The physico-chemical behaviour of this glass is covered in the previous section.

The second group includes the crystal and lead crystal glasses, in which most of the calcium is replaced by barium, zinc or lead, while sodium is in part replaced by potassium. The increased refractive index which results gives the glass a high degree of brilliance which can be further enhanced by cutting. The term *crystal glass* has been adopted due to its resemblance to precious stones. For several years the European Common Market countries have had legislation governing the quality criteria for crystal and lead crystal glass as follows:

1. *Crystal glass* must contain a minimum of 10% lead oxide (PbO), barium oxide (BaO), potassium oxide (K_2O), or zinc oxide (ZnO), alone or in combination, and its density (d) must be at least 2.45 g/cm^3. Its refractive index (n_D) must be at least 1.520 based on the yellow sodium-D spectral line. If there is no ZnO, however, the density must be at least 2.40 g/cm^3 and the Vickers surface hardness must be 550 ± 20.

2. *Pressed lead crystal* contains at least 18% PbO, d $\geq$2.70 g/cm^3, and $n_D \geq 1.520$. This designation may only be used in the Federal Republic of Germany.
3. *Lead crystal* contains a minimum 24% PbO, d $\geq$2.90, and $n_D \geq 1.545$.
4. *Full lead crystal* contains at least 30% PbO, d ≥ 3.00, and $n_D \geq 1.545$.

Density and refractive index were chosen as testing qualities because the glass does not have to be broken in the laboratory in order to measure them. Naturally the value of these glasses is not dependent upon their composition alone, rather it depends to a great extent on the finishing work involved and the standard of design. But lead crystal is generally considered to be especially valuable. In addition, as *long glasses* (see General Characteristics, Chap. 2), the lead glasses are easily shaped and cut, and can be acid polished. (See Finishing, below.) While their hydrolytic resistance is only slightly higher than that of soda-lime glass, they have excellent dishwasher resistance characteristics. The thermal coefficients of expansion lie between 7.5 and 9.0 $\times$ 10^{-6}/K.

The third group includes machine-pressed or -blown tableware made of special glasses immediately after the melting process. Their compositions vary greatly from those previously mentioned. There are currently 3 popular types on the market.

Transparent Borosilicate Glass. The *Duran* (Schott) and *Pyrex* (Corning, USA) borosilicate glasses which were introduced many years ago, are further developments of Otto Schott's *Geraeteglas* (apparatus glass), which was introduced into the household as *Jenaer Glas* during the 1920s. Dishes, baking molds, teapots, glasses and cups, coffee-maker jugs, baby feeding bottles, etc. are well-known products mass-produced at the melting tank from this type of glass. (See Figs. 5-16 and 5-17.) Thin-walled glasses are blown in hinged molds; thick-walled articles are pressed in a mold by a plunger. In the latter process, the hot glass is quenched with cold air in order to produce compressive stress in the glass surface (thermal toughening, see

Fig. 5-16. Tea service made of borosilicate glass.

Fig. 5-17. Pressed dishes made of borosilicate glass.

Safety Glass, Chap. 4). The firm's logo or trademark is also pressed in the bottom, or it is silk-screened and fired in as an after-process. Since the borosilicate glasses play an important role in technology due to their chemical durability and low thermal expansion, they are discussed in more detail under Specialty Glasses in Chapter 6.

White Opaque Glass (Translucent Glass, Opal Glass). Opaque (not transparent) glass is identified by an inhomogeneous composition, in which a microscopically fine distribution of at least two different types of material (phases) are present. Such glasses are formed when either one phase (such as tin oxide) does not dissolve during the melting process or it crystallizes out during cooling (fluorides); or the melt separates into two phases which, although glassy, exhibit strong light dispersing effects due to different refractive indices (as opposed to opaque tinted flat glass—see Opaque Glass, Chap. 4—which absorbs light). These glasses appear white or slightly tinted, according to the composition. (See Fig. 5-18.) The use of fluorides, which partially volatilize into the atmosphere, is hardly practiced anymore due to environmental laws. Therefore, such opacifying agents have largely been replaced by phosphates (primarily aluminum phosphate in combination with barium carbonate). The resulting white colored opal glasses, which are usually decorated as tableware, have a relatively low thermal expansion ($\alpha \approx 4.5 \cdot 10^{-6}/K$).

Glass Ceramics.   The principles of the manufacture of glass ceramics are covered extensively in Chapter 6 under Glass Ceramics. The products in this group, which were first brought out under the names *Pyroceram* (opaque) by Corning and *Jena 2000* (transparent) by Schott, have coefficients of thermal expansion which are close to zero. Therefore, they can withstand great temperature shocks. For example, they can be taken out of the freezer and immediately placed on a red hot stove. The opaque glass ceramics are optically closely related to the porcelains due to the fast growth of microcrystals during the ceramic-making process. (See Fig. 5-19.)

While bowls and dishes were previously exclusively pressed, new processes allow glass ceramic melts (normally difficult to shape) to be blown. This has allowed thin-walled glass ceramic articles to be introduced for household use and has extended the application of decorated temperature and thermal shock resistant articles. The same material, either dark-colored or white, is used for gas or electric stove tops.

Fig. 5-18. Decorated opal glass.

Fig. 5-19. Schott-Zwiesel's *Ceradur* glass ceramic pot on a *Ceran* glass ceramic cooking surface.

OTHER HOLLOWWARE

The technical applications of hollowware are so numerous that it is impossible to describe them all. For this reason, only several of the most common uses will be discussed here, and others will be mentioned in Chapter 6 under Special Glasses.

Hollow Structural Glass

Glass building blocks, "concrete" glasses, and glass roofing tiles comprise the hollow structural glass sector. They are manufactured by pressing. Rectangular glass building blocks are made by fusing two pressed halves together, whereby the pressure of the encased air is greatly reduced upon cooling. This creates good thermal and acoustic insulating characteristics (up to 40 decibels and more). The exposed surfaces can also have ornamental and light-scattering designs, colored decorations, or solar protective coatings. There are also panel shaped solid glass building blocks. Using suitable profiling and cementing materials, the individual building blocks are built into walls or permanent windows. They do transmit light but do not allow undistorted vision. They may not be used for load-bearing applications.

Concrete glasses, which are also pressed solid or hollow, are used in the manufacture of *glass-crete*, made of glass, steel, and concrete. These are used for cladding light shafts and can be walked upon and, in some cases, even driven upon. Glass roofing tiles are shaped like common clay tiles and must be strong enough to be walked upon (for cleaning) and to be hailstone-resistant. Concrete glass windows of a purely decorative nature have already been discussed in Chapter 4.

Lighting Glass

This glass should not be confused with lamp glass (Lamp Glasses, Chap. 6) which is mostly made from special glasses. Rather it is primarily used for all kinds of lighting fittings, including headlights. In addition to hollowware, a considerable amount of pressed glass parts are also used for this purpose. Like tableware, most lighting glass is machine-made soda-lime glass. To create brilliance, a great deal of crystal and lead crystal is also used (this is not covered under the crystal glass legislation). Also, opal glass is used as well as borosilicate glass for high-temperature lights (headlights, diazo printing equipment, oil lamps). Flashed opal glass is also used. (See Flashing, Chap. 5.) The manufacture of flashed opal glass has already been described in Chapter 4. The high degree of light diffusion of these glasses results in a certain amount of light loss; but it provides even, glare-free illumination. Colored glass is also flashed for lighting purposes. Lighting glass is supplied as desired by the designer to the light manufacturers for electrical fitting and completion. Glass factories also produce standard articles which can themselves be made into finished lights.

Laboratory Glass and Medical Hollowware

These comprise a large group of glass products which are used in many different areas, beginning in the laboratory and including industry, universities, standards bureaus, clinics, and chemical apparatus. A large and still growing portion is made from special glasses, covered in Chapter 6.

FINISHING OF HOLLOWWARE

In a strict sense, the finishing of hollowware pertains to the processing of the outer surface of finished glass which comes directly from the glass plant. As we have seen, this is also the case with flat glass.

In actual practice, the term is used much more broadly. It is divided into two categories: finishing in the hot state at the furnace and finishing in the cold state. Finishing at the furnace can only occur in a glassworks following shaping. Cold finishing pertains to processes which occur in the factory after the glass has been formed or to finishing companies which purchase raw glass from a glass manufacturer. But it can also entail reheating of the glass.

Finishing in the Hot State

The techniques involved here are of prime importance for tableware and lighting glass. Not all of them are economically significant; many have only historical importance or are no longer feasible. However, it is worthwhile discussing older methods of finishing glass as many of the processes are still being used.

Optically Blown Glass.   The light refraction of the glass can be emphasized through a wavy surface. To do this, the small parison is blown in a cylindrical, corrugated hollow mold. The surface design is maintained during subsequent processing, and it expands to cover the entire article being made. This is how longitudinally striped cylinder lenses are made. If they are twisted, they form a spiral.

Application of Filaments and Drops.   Molten glass which is drawn into thin filaments is arranged on the hot finished glass article. Depending on the arrangement of the filaments, this results in a spiral, ring, free-form, or picture design. (See Fig. 5-20.) Instead of filaments, individual droplets of glass called *nubs* can be applied. Pointed nubs hanging downward are called *snouts*.

Flashing.   Dipping a colorless glass parison into colored glass, followed by blowing, produces a so-called external flash. If the process begins with a colored parison, it is called an internal flash. (See Fig. 5-21.) A combination of both results in a double flash. Subsequent grinding or etching through the outer layer results in interesting effects caused by the contrasting colors.

Mosaic Glass.   Multicolored pieces of glass are distributed over a marble or iron plate. Over this the glass-blower rolls a hot, clear or colored blown glass to which the particles of glass immediately adhere. Then the entire object is overlaid with clear glass and finished in the usual way.

Fig. 5-20. Vase with filament application.

Fig. 5-21. Blown olive colored internal flash with a gold brown band.

***Millefiore* Glass (Thousand Flower Glass).**   Multicolored glass rods are bundled together, fused, and sliced into thin pieces. Further processing is similar to that for mosaic glass. (See Fig. 5-22.)

Filament Inlay.   A viscous filament of glass is applied to the hot glass and pressed into the surface by rolling back and forth on a plate made of marble or iron. Comblike tools can also be used to apply several filaments at the same time, resulting in a "combed" design. The color of the filaments must be different from the base glass; otherwise, they would not be visible. There is another process in which thin rods, usually made of opal glass, are applied along with colorless rods to the inside of a clay tube, and clear glass is then blown inside of it. The rods fuse with the main body of glass and result in a vertically striped piece of blown glassware, which can be made into a spiral by reheating and twisting. The entire item is then covered with a layer of clear glass.

Fig. 5-22. *Millefiore* bottle.

Air Bubbles.   If needles or nails are stuck into the surface of a soft glass, a small hollow area remains when they are removed, and the surface seals itself again. Each puncture leaves a bubble in the glass. (See Fig. 5-23.) The same effect can be achieved with molds having needles inside. Hollow stems in stemware are produced in a similar manner. They can be several centimeters long and can also be twisted.

Ice or Crackle Glass (*Craquelé*).   The parison is rolled in moist sawdust or covered with coarse sand to damage the surface. Then it is dipped in water. The quenching causes the surface to crack without destroying the glass. The blown glass is then covered with a fresh layer of glass or is reheated until the cracks fuse together somewhat, and the glass maintains its stability. (See Fig. 5-20.)

Foundry Ice.   The hot glass body is rolled over grainy glass powder, or the powder is sprinkled on it. A more or less rough surface results, depending on the size of the grains used.

Fig. 5-23. Tableware with an encapsulated bubble.

Agate Glass.   A marblelike glass structure can be created by mixing different colored, molten glasses. The resultant striped pattern is named after the semiprecious gemstone which it resembles. (See Fig. 5-24.)

Flame Coating with Metals.   Glass forms a secure bond with metals (Cu, Al) in the hot state (above 250° C). To a certain extent, the glass and the metal interfuse. The higher the temperature of the glass, the better the adhesion.

A flame gun is used to spray the metal on the hot glass (usually done as the glass comes out of the furnace to save energy; otherwise, the glass must be annealed and then reheated). In this process, a metal wire or metallic powder passes through the center of an oxyacetylene flame and is melted by the high flame temperature. It is "shot" with a large amount of kinetic energy onto the surface of the glass, where it bonds itself chemically to the glass in the area of contact. The surface becomes metallic. It can be altered or finished by brushing or buffing. Glass treated in this way is used for lighting, double-glazing and other applications.

Finishing in the Cold State—Removal Processes

After the finished hollowware has emerged from the annealing lehr, and after other hot finishing processes have been completed, such as removing the blowing cap (excess glass) in tableware and flaming the edges, it can be further finished by different techniques.

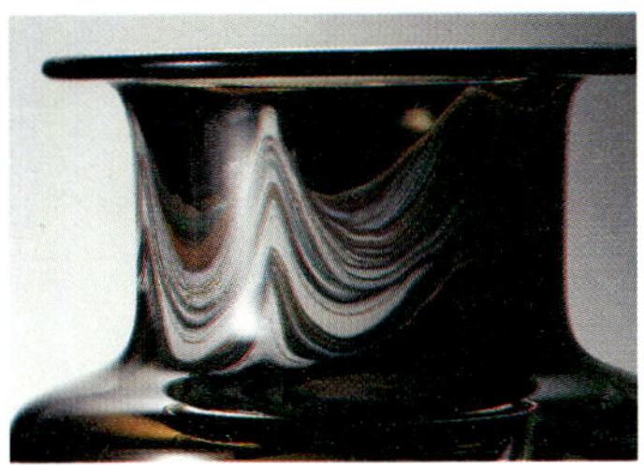

Fig. 5-24. Marbled vase (agate glass).

Grinding.   This is the most common means of finishing or decorating hollowware articles. (See Fig. 5-25.) Horizontal and vertical grinding wheels are used in this process. If the grinding surface of the wheel is flat, *flat or thumbnail grinding* results. (See Fig. 5-26.) If the wheel is rounded in a convex shape, the result is called *hollow grinding*. V-shaped grinding wheel surfaces produce wedge cuts. The grinder needs small wheels to make circular cuts, called *spheres*. If the grinding tools are somewhat larger, *olive cuts* are obtained.

Grinding Patterns are composed of several spherical or otherwise shaped areas. Wedge grinding can also be added. Combinations of grinding styles allow different decorative motifs to be ground on the glass, such as flowers. Usually the grinding is done in two or three steps which are performed by different workers. First, the pattern is rough-ground into the glass using an iron wheel and abrasives (such as sand, carborundum [SiC]) which remove most of the glass. Finer grinding follows. In addition to manual grinding processes, grinding machinery has been in use for a number of years.

Fig. 5-25. Creation of a ground pattern on table glass.

Fig. 5-26. Table glasses with thumbnail cut (left), diamond cut (center), notch cut (right).

Polishing.   When the glasses leave the grinding shop, the decorations are still rough-surfaced. In order to achieve the required high degree of brilliance, one of three additional processes can be used.

Acid polishing is conducted by placing the glasses in wire baskets, then dipping them in a mixture of hydrofluoric acid, sulfuric acid, and hot water, followed by rinsing with clean water. This process lasts up to 30 seconds and is repeated several times. Large batches of decorated glasses may be polished quickly in this manner.

Mechanical polishing is usually done with poplar wood wheels using a polishing agent (as a rule, polishing rouge, a fine-powdered iron oxide).

Lastly, there is *fire polishing*. In this, the glass is heated to 500°-700° C, whereby the plastic, viscous glass surface shrinks and becomes smooth as a result of surface tension.

Engraving.   Fine glass cutting is a refined method of grinding. Small copper wheels with diameters of 2-100 mm rotate on a horizontal axis. A grinding agent (emery in linseed oil) drips onto the wheels and helps remove glass stock from the glass article when it is held against the wheel from beneath. In raised or cameo cutting, the surrounding background is removed and a picture remains. The more popular deep cut leaves the glass surface intact but cuts the design into it. Line engraving is done like deep cutting and results in fine-lined designs without cut areas.

Etching.   This is based on the principle of acidic corrosion of glass. Hydrofluoric acid is most often used. Sometimes hydrofluoric acid vapors or baths of hydrofluoric acid salts are used. This depends on whether deep etchings are to be made in the glass or whether the surface should be matted only. The expert speaks of *deep baths* and *matt baths*. The matting (or frosting) of finished glass articles is important in lighting glass. Flashed opal glasses can be smooth or frosted for example. Because of the danger associated with hydrofluoric acid vapors and splashes, use of the etching process is avoided where possible. In many cases, it can be replaced by sand blasting.

Etched Designs   are made by coating glass with wax into which the desired pattern is scribed, either manually or mechanically, using a pantograph which reduces the size of the model, or with a Guillochier machine which can process 24 or more glasses. When liquid acid comes in contact with the object, the wax covered portions of the glass are not affected. Patterns and marks can also be applied directly onto the surface by using either etching ink with gold or platinum nibs, or rubber stamps. Graduation marks are applied in this manner.

Sand Blasting.   Deep-cut and matt surfaces can be produced by this method. The glass is protected by lacquer coatings or small rubber masks from the sand or corundum grit (Al_2O_3). As in the etching process, the covered surfaces of the glass remain untouched. Decorative patterns formed by sand blasting are well-suited for vases, large bowls, and other types of art glass.

Surface Coating Processes

Painting.   Enamel colors, which are fired in at 550°-650° C, are used for this. Basically, they are nothing more than easily fusible, pulverized colored glass mixed in a liquid (such as turpentine). Such colors are called *enamels* and are either opaque (containing opacifiers) or translucent. Drinking glasses, pitchers, vases, and lamp bases can be decorated by painting. *Black lead* is also an enamel, consisting of lead glass with additions of iron and copper. It was used in glass painting during the Middle Ages.

Color Staining.   Color staining occurs mainly with compounds of copper and silver. They are mixed with kaolin or ocher, stirred into a liquid, and applied to cold glass. Upon heating the glass almost to the transformation temperature, an ion exchange takes place between the alkali in the glass and the silver or copper ions. There are ruby stains and yellow stains. Black tints can be produced after immigration of silver by hot reducing gases such as hydrogen. Stained hollowware, such as vases or goblets, can also be engraved or ground.

Painting with Precious Metals.   Silver, gold, or platinum mixed with certain substances can be painted and fired into glass. After firing, the metal can than be polished to its original luster. This technique is most often performed on the rim of expensive drinking glasses. If an etched design is first applied to the glass, and silver, gold, or platinum is brushed over it and polished after firing, the cavities of the etched design remain matt.

Rubber Stamps.   These are suitable for application of single colors. The color is then fired in.

Steel Stamping.   Chiefly single color designs or drawings are manually or mechanically engraved in a steel plate or applied photographically. After the plate is colored, the picture is transferred onto pliable paper which is then pressed onto the glass. The design is then fired in.

Iridescent Glass.   Shimmering colors can be obtained if the glass is heated to at least 400° C, and vapors of solutions (e.g., chlorides) containing multivalent metals combined with moist carrier gases are directed over the glass surface using jets. The compounds decompose on the molten glass and are broken down into oxides. The colors appearing mainly in reflected light result from optical interference of the lightwaves when the thickness of the oxide layer is at least one-fourth of the wavelength and its refractive index is substantially higher than that of the glass. Among others the oxides of tin, titanium, and tungsten are used. As mentioned previously, similar effects are also obtained by using a vacuum-depositing technique in the production of glass jewelry.

Decals (Transfers).   Multicolored negative pictures or letters are first printed on a special carrier paper. They are then covered with another protective layer of paper. After removing the protective paper, the carrier paper is pressed onto the glass. After rinsing with water, the carrier paper covering can be removed. The transfer picture remains on the glass, and the colors are fired in. (See Fig. 5-27.) Promotional glasses, which are usually ordered in small quantities can, for example, be decorated this way.

Silk Screening.   This is a modern process for mass production. Photographically produced stencils are used in the printing process, with a separate stencil being used for each color. The printing machines operate with great precision so that even when several separate color printings are necessary, the register of each color is exact. After printing, the glasses pass through a firing furnace. The logos and names on beer glasses, name-brand beverage bottles, and name-brand jars for preserved goods are usually produced in a silk screen process.

Fig. 5-27. Example of decal decor.

6. Special Glasses and Their Uses

Unlike flat glass, fiberglass and hollowware, special glass is not identified by its appearance. The deciding factor is its use, and the properties of special glasses must be adapted to these applications. This is made possible by choosing suitable compositions which in turn requires intensive scientific research which few glass companies in the world can support. The result is a range of special glasses with high chemical and thermal durability and with a variety of optical, electrochemical or special technological properties. These glasses are used in such fields as chemistry, pharmacy, electrotechnology, electronics, apparatus and instrument construction, optics, illumination engineering, household appliances, certain sectors of the construction industry and in other technical applications.

FUSED SILICA (FUSED QUARTZ OR QUARTZ GLASS)

Of the single-component glasses, only the SiO_2 glass (quartz glass or fused silica) has technical importance. This is primarily because of its very low thermal expansion ($\alpha \approx 0.5 \times 10^{-6}/K$), its high thermal stability (up to almost 1000° C), and its extremely high UV-transmission. (See Table 6-1, glass #1.) Because the glass must be melted at temperatures above 2000° C, it is expensive and difficult to produce. A less expensive substitute is vitreous fused silica, which is a quartz glass melted at lower temperatures. Because it is not thoroughly refined, it is laced with small bubbles and not transparent.

In another process for making clear quartz glass, phase separable alkali-borosilicate glasses of low melting point are utilized. When the glass is thermally treated at around 600° C it separates into two phases. The alkali-borate rich phase may be leached out with acids, leaving behind open pores of controllable microscopic sizes.

By heating, the high silica (approximately 96%) phase which remains can be transformed into a clear glass. Known as *Vycor*, it has properties similar to pure fused silica (glass #2 in Table 6-1). For optical components such as UV light guides (Glass Fiber Optics, below), high-purity fused silica is obtained by pyrolytic decomposition of gas-forming silicon halogen compounds ($SiCl_4$).

Table 6-1. Examples of Special Glasses Used in Electrotechnology and Lamp Manufacturing.

Glass Number		1	2	3	4	5	6	7	8	9	10
Manufacturer		Osram	Corning	Schott	Schott	Schott	Schott	Osram	Schott	Osram	Corning
Code Number		452	7913	8487	8409	2877	8250	125	8531	713	7251
		(Fused Silica)	(Vycor)								
Transformation Temperature T_g (°C)		$\approx$1100	$\approx$1050	523	730	565	495	435	435	530	543
Temperatures (in °C)	10^{13}	1150	1020	535	730	570	507	429	430	550	544
at Viscosity	$10^{7.6}$ (dPas)	1650	1530	760	950	790	715	635	585	787	780
	10^4	—	—	1135	1235	1170	1060	1000	818	1224	1167
Coefficient of Thermal Expansion											
α_{20-300}°C (10^{-6}/K)		0.54	0.75	4.0	4.1	4.9	5.0	9.8	9.1	4.4	3.7
Density (g/cm³)		2.21	2.18	2.25	2.56	2.4	2.28	2.86	4.38	2.27	2.26
Elasticity Modulus E (10^3N/mm²)		66	68	64	90	75	64	—	52	59	65
Specific Electrical Resistance δ (Ω cm)*	log δ at 250°C	—	9.7	8.3	12.3	6.9	10.3	8.7	11.4	8.3	8.1
	350°C	—	8.1	6.9	10.5	5.7	8.5	7.1	9.8	6.9	6.6
Temperature at $\delta = 10^8$ Ω cm (°C)		510	358	275	535	195	384	280	448	278	—
Loss Factor† at 1 MHz ($10^4 \cdot \tan \delta$)		—	4	36	23	85	22	—	9	—	45
Composition (Weight %):											
	SiO_2	99.9	96	75.1	52.3	75.6	68.7	61.4	34.3	72.9	78
	Al_2O_3	0.005	< 0.3	1.3	21.8	4.5	3.0	2.0	—	4.5	2
	B_2O_3	—	< 3.5	16.7	1.8	8.8	18.6	—	—	14.5	15
	Na_2O	< 0.001	< 0.03	4.3	0.3	6.6	0.8	6.7	—	3.5	5
	K_2O	< 0.001	—	1.4	0.1	—	7.5	7.7	5.6	2.4	—
	MgO	—	—	0.4	7.2	—	Li_2O:0.6	—	—	—	—
	CaO	< 0.001	—	0.7	7.2	0.3	—	—	—	—	—
	BaO	—	—	—	1.9	3.9	—	—	—	1.2	—
	P_2O_5	—	—	—	7.0	—	ZnO:0.6	—	—	—	—
	PbO	—	—	—	—	—	—	21.7	59.6	—	—

*The logarithm of δ is listed instead of δ (log δ = 8.3, instead of $\delta = 10^{8.3}$).

†The phase angle between current and voltage is 90° for an ideal dielectric material; at a real 90° − δ; the (very small) angle δ is the loss angle, the relationship of the actual to idle power is the loss factor tan δ.

By adding 7%-10% by weight of titanium oxide to the SiO_2, α can be lowered to slightly negative values. Glasses with this composition were used in the United States as mirror substrates for telescopes, but more recently are being replaced by glass ceramics. (See Glass Ceramics, below.)

Leached SiO_2 glasses of the Vycor type having defined pore sizes are suitable for membranes in ultrafiltration and dialysis (e.g., for separating oil emulsions in water), and for anchoring biologically active materials, such as enzymes, in the field of nutritional biology.

BOROSILICATE GLASSES FOR INDUSTRIAL AND LABORATORY USE

Much mention has already been made of the many uses of the borosilicate group of glasses. A broad description of the composition and properties of the main types of borosilicate glasses was presented in Chapter 2, and their importance in household and catering glassware was mentioned in Chapter 5. The following chapters will discuss the purposes which these glasses serve in many technical areas—the chemical, pharmaceutical, and electrotechnology industries in particular. Technical data and the most important applicative properties of the main representative of this group, the Schott glass *Duran*, are listed below. The American glass, *Pyrex*, has almost identical behavior.

Physical and Chemical Properties of *DURAN* Glass (Schott)

Density	2.23 g/cm³
Linear thermal expansion	3.2-3.3 × 10^{-6}/K
Modulus of elasticity	6 × 10^4 N/mm²
Tensile strength (fire polished)	$\cong$90 N/mm²
Calculated continuous load factor	6 N/mm²
Refractive index (n_D)	1.473
Electrical volume resistance at 250°C	10^8 Ω cm
Transformation temperature (T_g)	530° C
Softening point	815° C
	water, according to DIN 12111 (5 classes) : 1
Chemical resistance to attack by: (best resistance is Class 1)	acids, according to DIN 12116 (4 classes) : 1
	alkalis, according to DIN 52322 (3 classes) : 2

In addition to being almost insensitive to temperature shock the borosilicate glasses exhibit the following advantageous qualities: will not deform until approaching 550° C, no gel layer formation on the surface (hence low resistance to flow and little tendency to separate), the glass yields no metals when in contact with liquids, does not act as a catalyst, and is resistant to radioactive radiation. (See Fig. 6-1.) As a *long glass* (General Characteristics, Chap. 2), it is suitable for hot processing even for complicated component shapes and apparatus.

Laboratory Equipment

Only a few characteristic examples can be cited here of the broad variety of borosilicate glass apparatus used in research, development and test laboratories of all kinds, and in process technology. Beginning with test tubes, laboratory equipment includes beakers and flasks, volumetric glassware (such as graduated measuring cylinders, pipettes, and burettes), filtering devices, gas washing bottles and reaction vessels, distillation equipment, condensers and heat exchangers, as well as stopcocks and valves for liquids and gases. Special glass types are used to make thermometers; one alkali-free glass enables precise temperature readings of up to 620° C to be made. According to the type of laboratory, additional specialized apparatus and measuring equipment made of glass are used to test, convert, or separate materials. Examples of this include apparatus for gas analysis, titration, liquid and gas flowmeters, capillary viscometers for liquid viscosity determination by flowrate, and equipment for molecular distillation used in the production of vitamin extracts or essential oils and fragrances. Additional process equipment, such as stirrers, pumps, high-vacuum and diffusion pumps, and rotary evaporators also belong to this product group (Fig. 6-2); along with different types of easily fitted connectors for linking equipment parts, which utilize tapered and flat-flange connections or glass threads, assuring liquid- and gas-tight seals in all cases.

Glass Process Plant

This is an extension of laboratory equipment; only the amount of material to be treated or conveyed is several times larger. Such glass

Fig. 6-1. Laboratory Glass.

systems are generally built as pilot plants before being constructed for full-scale production.

By using standard interchangeable parts, it is possible to construct complicated facilities having many components, large dimensions, long pipelines, and many side branches. Currently, borosilicate glass

Fig. 6-2. Chemical centrifugal pumps.

can be processed into pipes having diameters up to 300 mm and vessels with 500 liter useful capacity. (See Fig. 6-3.) Straight pipes and bends, valves, column sections, and heat exchangers are available in a range of nominal bores and modular lengths. Various types of glass pumps working on different principles are also used in such constructions.

Glass pipelines of any length for wastewater or gases can be installed inside and outside buildings in pipe shafts, channels, or even underground. To prevent excess strain, they are not installed rigidly, but suspended and connected flexibly. For pipes having thin flanges at each end, it is in many cases sufficient to use rubber sleeves and metal clamps.

In addition to spherical and cylindrical vessels, columns in sizes ranging up to 20 meters in height and 1 meter in diameter are used in chemical process systems. (See Fig. 6-4.) They are used for extraction and heat exchange between rising vapors and descending liquids. They usually also contain small pieces of glass tubing (*Raschig rings*)

Fig. 6-3. Tubular heat exchanger made of borosilicate glass.

Fig. 6-4. Chemical plant made of borosilicate glass.

which serve to increase both the surface area and time of contact of the flowing media. The throughput can be as much as several tons per hour.

Glass systems are used in the pharmaceutical industry for the extraction and distillation of substances and for solvent recovery. Distilleries use continuous fractionating columns for purifying and concentrating distillates. Glass pipelines are used extensively in all types of beverage production, in dairies, in the production of aromatics, and in special branches of the food processing industry. Textile finishing plants use them in dyeing equipment. In the electroplating industries, vacuum concentrators made of glass are used for recovering metals and chemicals from the electroplating baths. The dimensions of plant and pipeline components are extensively standardized (e.g., ISO and DIN) and these standards are constantly being added to. Components of different manufacture can therefore often be used together.

PHARMACEUTICAL GLASS

Many pharmaceutical preparations are packaged in glass containers such as ampoules, vials, and small bottles. (See Fig. 6-5.) While ampoules are always made from glass tubing, small bottles can be made either from tubing or from molten glass directly from the melting tank. The *pharmacopoeia** (European = Eur. AB; German = DAB) dictate that injectable substances may only be bottled in containers, the hydrolytic surface resistance of which meets the requirements of Glass Type I of the Eur. AB (Durability Class I) or Durability Group A of the DAB (in accordance with DIN 52329; or ISO 4802, in preparation). The *Fiolax* clear (also called *Fiolax* colorless or white) and the *Fiolax* brown made by Schott-Ruhrglas are borosilicate glasses which meet these requirements. These glasses contain less boric acid than *Duran* glass (see above) and more alkali oxides, along with several percent calcium oxide and barium oxide.

The coloring agents of the brown glass are iron oxide and titanium oxide. Both glasses are well-suited for pharmaceutical use due to their high chemical durability. (See table below.) In addition, they possess good glass processing characteristics and are highly suited to automatic machinery for processing glass tubing into containers having

Chemical and Physical Properties of Glass Tubing for Pharmaceutical Applications

	FIOLAX CLEAR	FIOLAX BROWN	ILLAX	AR-GLASS
Hydrolytic class	1	1	2	3
Surface resistance (Eur.AB)	I	I	I	III
Surface resistance (DAB)	A	A	B	C
Acid durability	1	1	2	1
Alkali durability	2	2	2	2
Coefficient of thermal expansion ($10^{-6}K^{-1}$)	4.9	5.4	7.6	9.0
Transformation temperature T_g (° C)	560	550	528	520
Working temperature Va (° C)	1160	1145	1058	1035
Density (g/cm³)	2.39	2.44	2.50	2.52

Pharmacopoeia are official collections of regulations concerning quality, dosage, testing, and storing of drugs. Most advanced countries publish them.

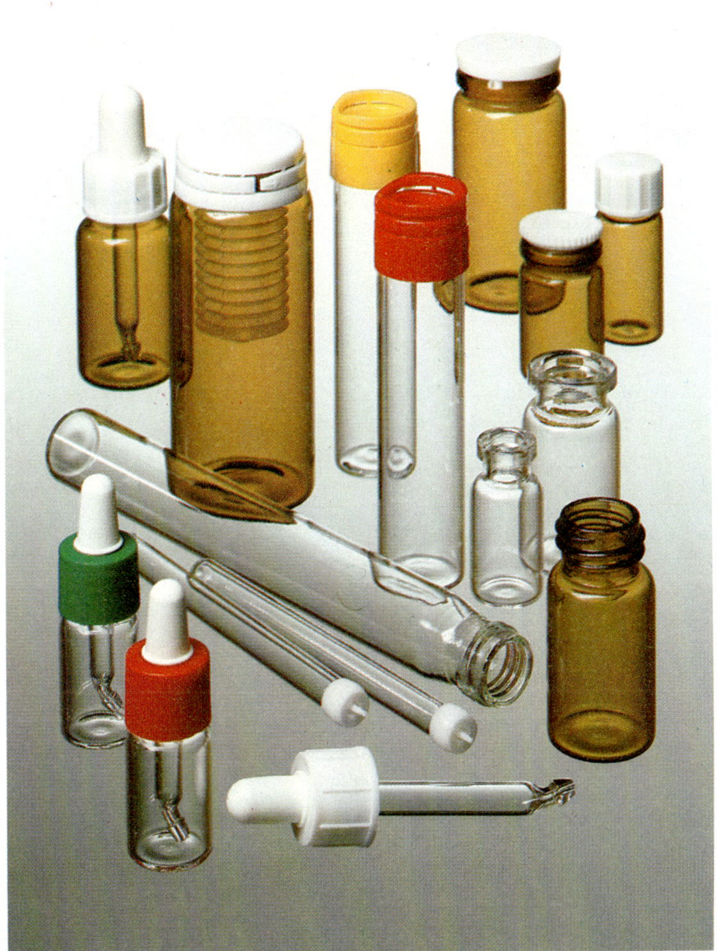

Fig. 6-5. Pharmaceutical bottles.

the close dimensional tolerances required by the very fast filling equipment used in the pharmaceutical industry.

The spectral transmission of the brown *Fiolax* glass, which is used to protect against sunlight, is shown in curve a in Fig. 6-6. Curve b

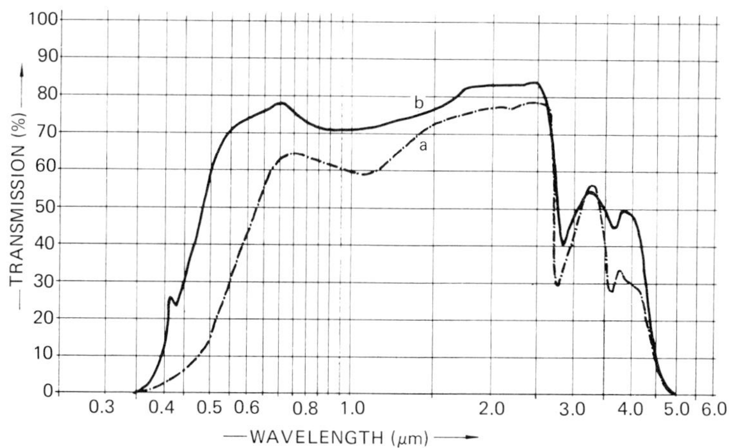

Fig. 6-6. Spectral transmission of *Fiolax* brown (a) and *Illax* (b) at 1 mm thickness.

represents *Illax* glass which is colored with iron oxide and manganese oxide for use with less sensitive preparations. It has a high degree of chemical durability, but it may not be used for injectable substances because it is a Class 2 glass in regard to hydrolytic resistance. It is used for drinking vials, containers for sensitive reagents, tubes for pills which are sensitive to light and air, medicine bottles, etc.

The last glass in this group is the *AR*-glass which is a colorless soda-lime glass containing only 1.5% boric acid. Since this glass is in hydrolytic class 3, it is mainly used for bottles for dry or water free (oily) preparations.

GLASSES FOR ELECTROTECHNOLOGY AND ELECTRONICS

The special glasses used in electrotechnology are characterized by excellent electrical insulation, low dielectric loss, gaseous impermeability, high absorption of certain radiation, and "tailor-made" thermal properties. The most important requirements concern absolute gas impermeability in evacuated tubes and the associated glass-to-metal seals. These seals require painstaking matching of the coefficients of thermal expansion (α) of glass and metal to ensure that they maintain impermeability and freedom from microcracks under all possible operating conditions. Such glasses are called *sealing glasses*.

In many cases, other special properties are also required. These include resistance to certain gases and vapors (in alkali vapor lamps), absence of alkali oxides, and low working temperatures.

Sealing Glasses

The different sealing glasses for glass-to-metal seals are grouped according to the thermal expansion ranges of the available sealing metals. When a direct seal between a glass and another material is not possible because of strain caused by excessive differences in α-values, transitional glasses must be used between the two components.

Tungsten.   Tungsten, which has the lowest expansion coefficient of the metals to be sealed, requires glasses with coefficients of thermal expansion (α) from 4.0 to 4.4 $\times$ 10^{-6}/K. Borosilicate glasses are suitable in such a case (Glass No. 3 in Table 6-1). They are especially well-suited for making high temperature incandescent and discharge lamps (flash tubes). Operating temperatures as high as 750° C can be attained with alkaline earth-alumino-silicate glasses (Glass No. 4), the transformation temperature of which can lie as much as 200° C above that of other multicomponent glasses. Since they are essentially alkali-free and therefore highly insulating glasses, they are suitable for use in halogen-tungsten lamps and also as carriers for high throughput surface resistors.

Molybdenum.   Molybdenum, which due to its high electrical conductivity has maintained its popularity against the Fe-Ni-Co substitute, is a classical sealing material like tungsten. There are a number of sealing glasses for molybdenum and Fe-Ni-Co alloys with α-values around 5 $\times$ 10^{-6}/K. The alkali-borosilicate glasses (Glass No. 5) are particularly suitable if the glass is not required to have special electrical qualities. If high insulating values must be maintained to around 300° C (such as internal glass parts or high power lamps), glasses having higher percentages of B_2O_3 and lower percentages of Na_2O are used.

Kovar.   Even higher percentages of B_2O_3 (17%-23%) are necessary if the transformation temperature of the glasses must be reduced below 510° C in order to fuse them with Kovar metals (e.g., 28% Ni, 18% Co, 54% Fe) in spite of the transition point in their expansion curve. This alloy is very important for glass-to-metal seals due to its extremely low thermal expansion and its high degree of corrosion resistance. Hence a number of glasses with special properties are designed for use with this alloy. These glasses have such special qualities as low X-ray absorption, high UV-transmission, and high electrical insulation (Glass No. 6).

Lead Glasses.   These fulfill a number of important functions in electrotechnology and electronics. Their favorable electrical qualities are based on the fact that the large, heavy lead ions block the electrically induced migration of the alkali ions within the glass network. Thus they inhibit conductivity and dielectric losses. Lead glasses (Glass No. 7) have importance as bases for incandescent and discharge lamps, for television tubes, and for many other types of vacuum tubes. They are compatible with numerous Ni-Fe-(Cr) alloys and copper clad wire.

In electronics, high-lead-content glasses are chiefly used for the encapsulation of diodes and other components, such as precision resistors and ceramic or tantalum condensers. (See Fig. 6-7.) A highly insulating, alkali-free lead glass (with about 60% PbO) has been developed for use in alkali-sensitive silicon diodes. It has a very low fusing point (Glass No. 8). In order to obtain a hermetic seal and hence provide chemical and mechanical protection for the components, sections of fine tubing are pushed over the components and are then sealed in a protective gas atmosphere using heating coils or continuous furnaces.

Infrared Absorbing Glasses.   In addition to using flame and heating coils to heat glass for hot shaping, IR heat transfer has also been introduced, usually using quartz-iodine lamps. Glasses containing several percent iron oxide (FeO) are used for this purpose. They ab-

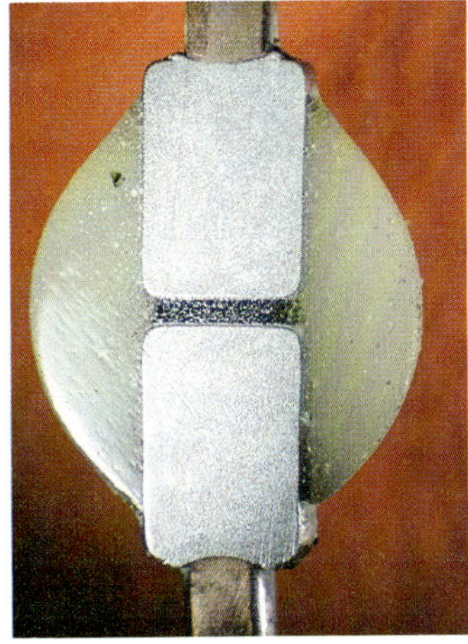

Fig. 6-7. Section of glass encapsulated silicon diode.

sorb infrared and red light, giving them a greenish tint. Unlike flame heating, hot shaping by means of radiation can be conducted in a neutral or reducing atmosphere. This is advantageous in the production of hermetically encapsulated electrical components filled with special gases.

The process is used chiefly in the manufacture of reed switches. (See Fig. 6-8.) They consist of two magnetically activated contact

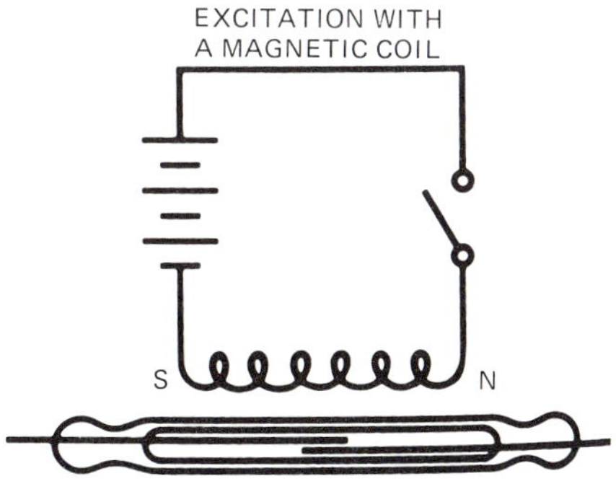

Fig. 6-8. Principle of a reed switch.

reeds which are hermetically sealed into the ends of the glass envelope. The glasses are also used as sealing glasses for Ni-Fe alloys.

Glasses for Television Tubes

In addition to the requirement that they be capable of fusion with metal electrodes carrying high voltage as well as able to withstand heating and control voltage in the tubes, X-ray absorption is also of prime concern in glass used for the production of television tubes. (See Fig. 6-9.) The glasses must have minimum amounts of heavy oxides in the glass (BaO, PbO, SrO), which almost completely prevent the escape of X-rays produced in the tubes. Glasses having different absorption characteristics must be used, according to the wall thickness of the tube parts (panel, funnel [cone], neck). There are also special requirements for the glass used in the neck regarding resistance to high voltage. The alkali-barium-silicate glasses are suitable for use in picture screens (panels), and the lead glasses are used in the funnel and neck, based on optimum shaping properties and the best possible glass quality (freedom from knots, cord, and stones).

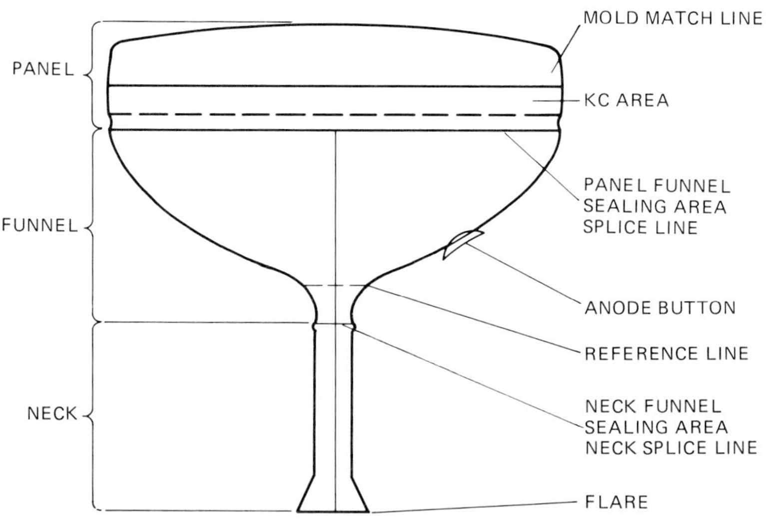

Fig. 6-9. Schematic construction of a television tube.

Glasses for X-Ray Tubes, Transmitting and Image-Intensifying Tubes

For decades, glasses with a high content of barium- or zinc-oxide were used to make X-ray tubes. They resulted in relatively high absorption losses for X-rays. Only with the advent of newer glasses developed by Schott, which contain elements with atomic numbers below 20 (atomic weights below 40), has satisfactory minimum X-ray absorption been made possible while still meeting the requirements for sealing to metals. They resemble glass number 6 in Table 6-1 and are compatible with molybdenum and kovar metals (e.g., *Vacon 10*). Since bubble and striae quality specifications are very stringent, they are melted in processes similar to those used for optical glass. (See Manufacture of Optical Glass, below.)

The same glasses are also used for image-intensifying and image-converter tubes, vidicons, mercury switches, and high power transmitting bulbs which require high insulating capability and low dielectric loss.

Glasses for Soldering and Passivation

Similar to metallic solders, glass solders have a very low melting temperature and can be used to solder all normal glasses, metals, or ceramics with as little exposure to excessive temperature as possible. When glass itself is a component to be soldered, the glass solder must flow and interfuse at temperatures well below that at which the glass to be soldered will deform. Generally this temperature is determined by the transformation temperature of the glass to be sealed. Such solder glasses generally flow and interfuse at temperatures between 450°-550° C. In order that these soldering glasses flow and set, the viscosity required at these soldering temperatures must lie around 10^4 to 10^6 P. (See footnote, p. 19.)

As is also the case with sealing glasses, the thermal expansion of soldering glasses is determined by the component to be soldered. Since the solder is normally mechanically weaker than the mating surface, it is desirable to put it into compression during the cooling cycle.

Glass solders usually come in powdered form (or sometimes in sintered preforms) with grain sizes of 60 μm or less. They are usually applied to the glass to be soldered in the form of a suspension or paste with a carrier liquid such as water or methanol. By using nitrocellulose dissolved in amyl acetate, the powder will adhere, even after the solution dries. The nitrocellulose binder evaporates before the sintering phase begins.

Solutions may be applied by spraying, silk screening, or extruding. After the mating surfaces come in contact, the soldering process is conducted using a suitable temperature-time program.

Glass soldering is usually done when complete fusion is not technically feasible. Examples include picture tubes for colored television containing many components, and flat numerical displays (liquid crystal and gas discharge displays).

Glassy Solders.   These do not noticeably crystallize during the soldering process (which lasts for several minutes to an hour). In other words, the viscosity is reversible during heating and cooling (thermoplastic glass solders). For this reason, mating surfaces joined by glass solder can be reset or separated when reheated.

Glass solders having these characteristics are generally lead-borate glasses with 60%-90% PbO by weight. To improve chemical resistance, they usually contain SiO_2 and Al_2O_3.

Crystallizing Solders.   These generally maintain their glassy character until they reach the soldering temperature. Due to separation of crystals, they transform into a glass-crystalline or ceramiclike body when the soldering temperature is maintained. This process results in an irreversible joint preventing relocation of the mating surfaces. This also holds true if the soldered articles are subsequently reheated to soldering temperatures. Such fused connections can only be separated by chemical dissolution of the solder layer.

Crystallizing solders chiefly differ from glass solders in their zinc oxide content (8%-25% by weight). If allowable soldering temperatures exceed 550° C (for ceramics or metals), zinc-borate and silicon-borate glasses (50%-65% ZnO, 0%-15% SiO_2, 20%-35% B_2O_3 by weight) can also be used.

Composite Glass Solders.   According to a general glass technological law (Soda-Lime Glasses, Chap. 2) the tendency to develop glass solders which have the lowest possible application temperatures introduces a corresponding and undesirable increase in thermal expansion. This effect is in fact smaller in crystallizing solders. However, a far more effective method entails the addition to the powdered glass solder of inert (nonreacting) fillers which have low or negative α-values. Included among such suitable materials are the minerals, zircon ($ZrSiO_4$) or β-eucryptite ($Li_2O \cdot Al_2O_3 \cdot 2SiO_2$), which are also useful in the manufacture of glass ceramics. (Compare with Glass Ceramics, below.) Such composite glass solders are usually used as stable glass solders. The amount of fillers which can be mixed into the batch is limited by the unavoidable lessening of flow capability during the soldering process.

Passivation Glasses.   These are glass solder-related zinc-silicoborate and lead-alumosilicate glasses. They are used for chemical and mechanical protection of semiconductor surfaces, especially silicon components. They are applied as a powder (grain size $= 10\text{-}20 \ \mu m$) and melted down as either a 10-50 μm thick cover layer or as a housing. Of utmost importance is the complete exclusion of alkalis (since these can disturb the semiconductor function), and the dielectric and mechanical resistance of the glass layer. Alkali-free composite glasses have proven effective in matching the thermal expansion of the silicon ($\alpha \ = \ 3.3 \ \cdot \ 10^{-6}/K$) at moderate melting temperatures ($700°\text{-}800°$ C).

Shielding from alkali immigration is usually also desirable if the semiconductors are to be bonded to glass substrates containing alkalis, as is the case in the transparent electrode layers in liquid crystal displays. To accomplish this, an SiO_2 layer is usually applied by dip-coating or spraying or by precipitation in gaseous reactions. Other oxides may also be used.

Sintered Glass Parts

Hot forming processes are neither economically feasible nor sufficiently precise for production of small components in large quantities, such as used in electronics. For a long time, such small components have been made from dry- or wet-milled glasses. Glass powder with average grain size of about 50 μm is mixed with plasticizing organic binding agents and pressed dry. When heated slowly, the organic substances burn off. Then the sintering process begins with additional heating to 600°-700° C, whereby the glass frit solidifies into a mechanically stable gastight glass body. Dimensional tolerances can be held within 2%. Such pressed granulates are made of many different types of special glass, especially the sealing glasses. (See Sealing Glasses, above.)

A primary application of sintered glass parts is in insulated bushings for electrical leads in hermetically sealed housings. (See Fig. 6-10.) The volumes of these can vary from a few cubic millimeters in semiconductor components, up to several cubic meters in nuclear power facilities. (See Fig. 6-11.) They usually consist of a metal ring in which a sintered glass component and one (or several) electrical

Fig. 6-10. Glass-metal bushings for transistors, crystals, and rectifiers.

Fig. 6-11. Glass-metal bushings in the nuclear powered ship, the *Otto Hahn*.

conductor(s) are fused. Other sintered glass products include insulators, precision distance sleeves, transistor bases, relays, quartz crystal
oscillators, etc.

Glasses for High-Voltage Insulators

These are in general pressed from soda-lime glass (sometimes containing PbO) and are usually thermally toughened to increase their
mechanical strength. (See Safety Glass, Chap. 4.) For voltages in
excess of 20,000 V, borosilicate glasses are preferred because of their
resistance to thermal shock and moisture. The electrical breakdown
level of the glasses is approximately 450,000 V/cm at 50 Hz. In competition with ceramic insulators, they can be used successfully in
overhead railway power lines and in overland high-voltage lines. The
insulators are suspended so that the individual glass elements are not
subjected to any tensile stress.

Ultrasonic Delay Lines

In the transmission of electrical signals, it is often necessary to have
precisely defined delays in signal transfer. This is difficult to achieve

by purely electrical means. The problem arises, for example, in the creation of colored pictures in television tubes. The three electron beam systems for the red, green, and blue color components scan the screen in lines, thereby determining the brightness and color composition of each point on the screen. The beams must be synchronized in sequence at 64 microsecond intervals (6.4×10^{-5} seconds), which is the time needed to traverse one line. This delay is very long compared to transfer times of electric signals and is achieved by transforming the electric signal into an ultrasonic signal through a small glass plate with several parallel side surfaces. The signal is received at one of these surfaces, and an electromechanical transformer converts it. After being reflected back and forth several times in the glass, it is once more transformed into an electrical signal at the exit surface. Several special lead glasses can fulfill the requirements of low ultrasonic loss and minimum temperature dependence of the sound wave propogation better than any other materials. Glass delay lines are also used in computers and radar. If temperature independence is not very important, quartz glass (Fused Silica, above) can be used because of its low attenuation of sound waves.

Electron Conductive Glasses

The previous sections and Table 6-1 have demonstrated that the specific electrical resistance of glasses $(\rho)^*$ is higher when they contain fewer alkali ions. At normal temperatures, values lie between $10^{12}\,\Omega\text{cm}$ (very high alkali content glasses) and $10^{20}\,\Omega\text{cm}$ (fused silica). The volume conductivity (as distinguished from the surface conductivity which arises from adsorbed water) is based on the ability of these ions to migrate in an electrical field. This is very slight at room temperature, but increases quickly as temperature rises. Under sustained direct current, the mobile metallic ions can collect at the cathode and cause polarizing effects.

However, there are alkali-free glasses which have resistances of only 10^4 to $10^{10}\,\Omega\text{cm}$ at normal temperatures. They conduct electrons because they contain several multivalent elements. This is the case

---

*ρ = resistance of a cube with edge measurements of 1 cm.

with vanadium-phosphate glasses (V_2O_5 - P_2O_5), which contain both V^{4+}, and V^{5+} ions when there is an oxygen shortage. The electrical conductivity is a result of electrons jumping from one ion to another. Typical semiconductor effects are also present in the chalcogenide glasses. The elements sulfur (S), selenium (Se), and tellurium (Te) belong to the chalcogens. In combination with arsenic, antimony, germanium, and/or halides, low melting-temperature glasses can be made in the absence of oxygen. These are of great interest because of their infrared transmittance and their sudden change from conditions of low conductivity to high conductivity when certain threshold voltage values are exceeded. However, development in this field has not yet reached technical maturity.

Semiconducting oxide glasses are of particular interest in the construction of multipliers (secondary electron multipliers). For example, if about 500 volts per cm length are applied across the ends of a thin tube made of such a glass, a stream of electrons flows within the glass. Free electrons emanating from light beams of a photo-cathode and directed into the tube displace several secondary electrons when they hit the wall of the glass. These secondary electrons are accelerated by the field and in turn bring about the same effect. If this process is only repeated along the channel ten times, a million times more electrons meet at the receiver end (e.g. fluorescent screen) of the "microchannel" than enter at the arrival end.

This principle was further developed for image intensification and transformation of weak or invisible (infrared, UV, X-ray) radiation into visible light. The required large number of picture points is achieved by using a bundle of thin hollow glass fibers, the length of which is ten times the internal width. To make such hollow glass fibers, clad fibers (Glass Fiber Optics, below) with an acid-soluble glass as the core are drawn, fused into a bundle, and cut into sheets. The core glass is then leached out with acid. In order to prepare the resultant microchannels for generating secondary electrons, they can—if the glass itself is not a semiconductor—be coated with conducting oxides in gas reaction processes. The cladding glass is however usually a lead glass which forms its own conductive surface layer in hot reducing gases.

LAMP GLASSES

The most important requirements of a glass used in electric lamps are that it form a light-transmitting covering for the actual light source (such as the tungsten filament in incandescent lamps) and that it be impermeable to gases at high temperatures. The current is delivered by way of metallic conductors which are fused in the glass. A gas- and vacuum-tight seal is best achieved when the glass and the metal both have similar coefficients of expansion. (See Fig. 6-12.)

For *low power lamps*, copper clad wire is usually used to supply the current. This wire has an iron-nickel alloy core which is clad with a 20%-30% copper layer. The metal combination has a radial coefficient of expansion of 9.0 to 9.5 $\times$ 10^{-6}/K. Accordingly, soft glasses having α-values between 9 and 10 $\times$ 10^{-6}/K are required for the lamps.

The most widely used type of *incandescent lamp* is the everyday pear shape. The bulb is made of a simple soda-alkaline earth-silicate glass and is machine blown on high throughput automatic machines (e.g., ribbon machines). (See Shaping Hollowware, Chap. 5.) The electrical insulation of the lamp is too low at operating temperatures, so a highly insulating glass must be used as the base of the lamp carrying the sealed-in wires. Lead glasses having 20% to 30% PbO content are used for this purpose (glass No. 7 in Table 6-1). These glasses are drawn in tube form by Danner and Vello machines. (See Drawing Process, Chap. 5.) To produce internally frosted bulbs, the inside surfaces of the glass are etched with a mixture of hydrofluoric acid and fluorides to obtain the desired degree of surface roughness.

Quite a few other incandescent lamps are made by using the same process (for example, for vehicle, photographic, and microscope lights).

Fluorescent lamps are low pressure mercury-discharge lamps with a fluorescent luminous substance coating the inside of the tube. In a tube made of soda-alkaline earth-silicate glass, mercury vapor is excited for discharge and the resulting UV-radiation is transformed into visible light when it strikes the fluorescent material on the inside of the glass bulb. Due to the high UV-absorption of the fluorescent material and of the glass, no UV-light is emitted externally. Lead glass is also used to imbed the current leads at the ends of the lamp.

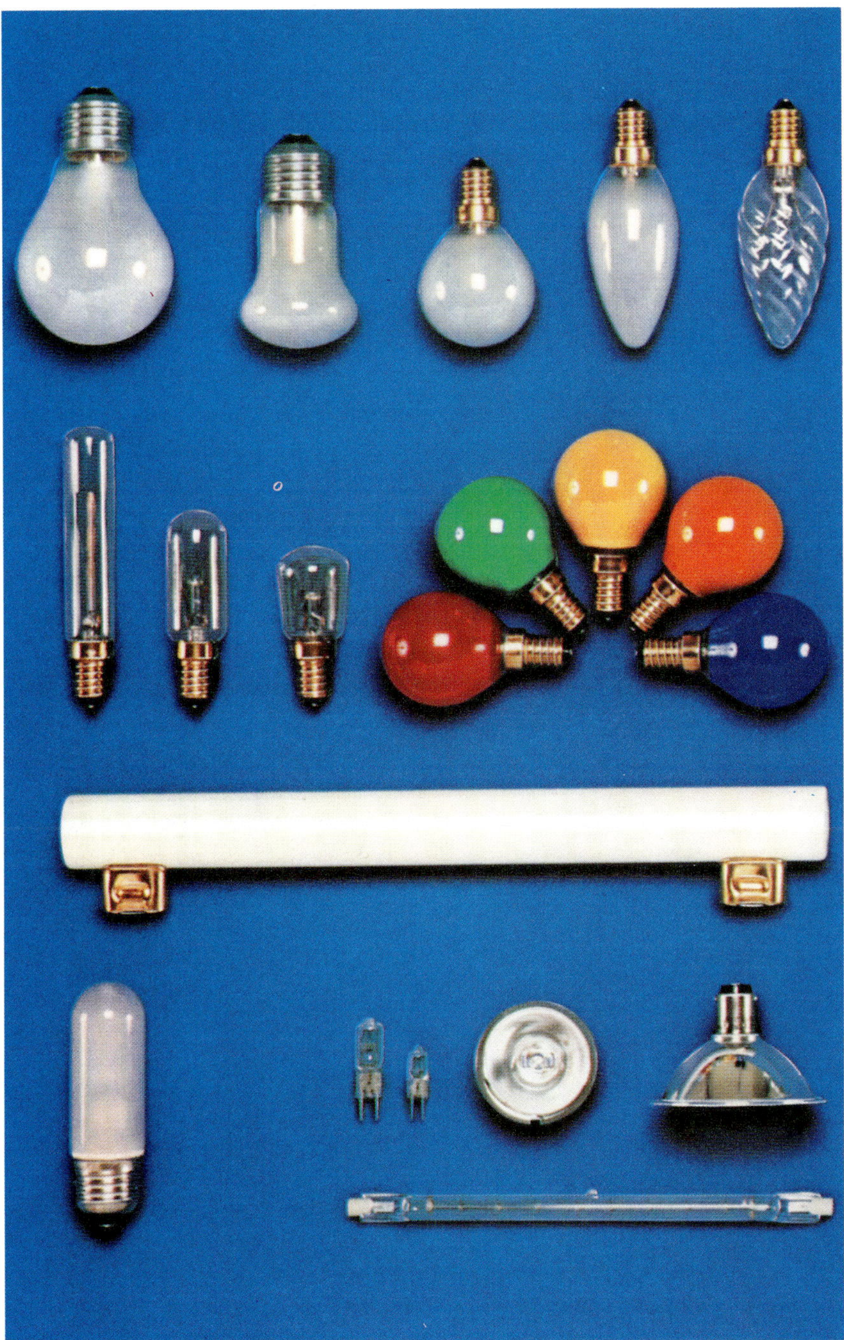

Fig. 6-12. Different types of lamps made by Osram.

The soft glasses do not have sufficient resistance to thermal shock to be used in high electrically or thermally loaded lamps. Hard glasses with coefficients of thermal expansion below $5 \times 10^{-6}/K$ are used (e.g., glass No. 9 in Table 6-1). Molybdenum, tungsten, and Fe-Ni-Co alloys are the conductive materials used in lamps within this expansion range.

A number of focused incandescent lamps, such as the CONCEN-TRA® or sealed beam lamps used in the United States, are made of pressed glass parts (glass No. 10 in Table 6-1). Both the reflector and the cover are made of borosilicate glass with 12%-15% B_2O_3. The reflector is silvered in the lamp factory (e.g., aluminum is vacuum-deposited on the reflector), the tungsten filament assembly is inserted, and the cover is fused to the reflector.

In *projection bulbs*, the bulb temperature exceeds both the annealing temperature and the softening point of the soft glasses and most borosilicate glasses. Bulbs must, therefore, be made of glasses which have transformation temperatures above 700° C. The glasses used are alkaline earth-alumino-silicate glasses (similar to glass No. 4 in Table 6-1) with 18%-24% Al_2O_3 and 5%-9% $B_2O_3 + P_2O_5$. The base is made of highly insulating borosilicate glasses, usually tungsten sealing glasses.

Halogen incandescent lamps have halogens added to their filler gases. The halogens inhibit the blackening of the surface of the glass bulb caused by the vaporized tungsten. Due to the higher bulb temperature required for the so-called halogen ring process, these lamps must be much smaller than normal incandescent bulbs of similar power. For this reason, quartz glass or Vycor (Fused Silica, above) are used. Because of the vast differences between the thermal expansions of quartz glass/Vycor and molybdenum ($5.4 \times 10^{-6}/K$), a vacuum-tight filament seal is not possible. The electrical contacts are therefore made of a 20-40 μm thick molybdenum foil which is pinched into the softened tube.

For *low power halogen incandescent lamps*, alkaline earth-alumino-silicate glasses (similar to glass No. 4 in Table 6-1) are also used. Their transformation temperatures lie between 700° and 800° C, and they are fused directly to the molybdenum filament. The glass tubes with few exceptions are machine drawn.

In *high pressure mercury discharge lamps*, the radiation is generated in a tube of high UV-transmitting quartz glass. A molybdenum film is again used as the electrical conductor. The discharge container is sealed inside an external bulb which is coated with fluorescent material, so that the UV-radiation is also converted into visible light. (See Fig. 6-13.)

Low power lamps have external bulbs made of soda-alkaline earth glass, and bases made of lead glass. Lamps above 250 watts have bulbs made of borosilicate glass and bases made of highly-insulating borosilicate glass.

The discharge containers of *special high pressure discharge lamps*, (e.g., xenon lamps) become very hot (between 1000°-1200° C). Therefore, quartz glass is used. Lamps having relatively low current loads are manufactured with molybdenum foils as electrical conductors. Those with higher current loads require tungsten rods. In order to overcome the high difference in thermal expansion of the quartz

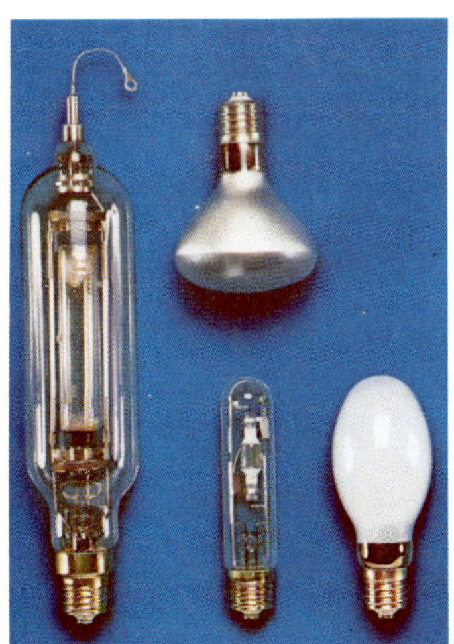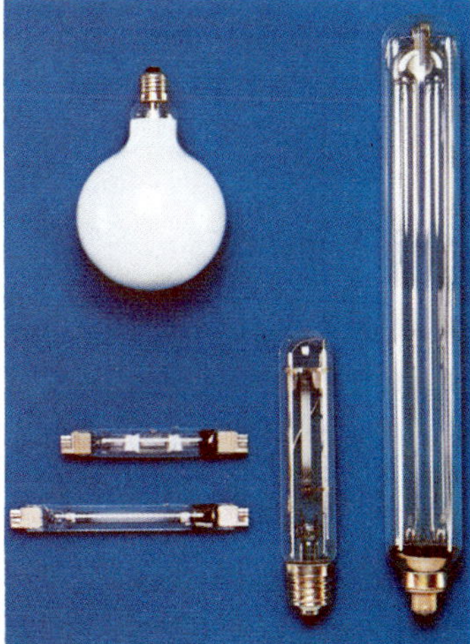

Fig. 6-13. Discharge lamps.

glass and the tungsten, two or three intermediate seal glasses with coefficients of expansion of 1.3, 1.8, and 2.3 $\times$ 10^{-6}/K are applied. The glass having the highest expansion is fused to the tungsten.

In *sodium discharge lamps*, sodium vapor is excited to the discharge point. The low pressure lamp emits an almost pure monochromatic yellow light. At high temperatures, sodium vapor destroys normal technical glasses. For this reason, special glasses have been developed, the most widely used of which are the barium-borate glasses which are low in silica content. Most of these glasses are difficult to work (short), or they are insufficiently resistant to weathering. Therefore, a two-layer (flashed) glass is used for the bulb. The soft glass tube (alkali-alkaline earth-silicate glass) is internally coated with a thin layer (50 to 100 μm) of the sodium-resistant glass. The flashed glass tubing is mechanically drawn.

The high pressure sodium lamp produces a yellowish-white light, and the discharge container heats up to about 1100° C. Thermal and chemical attack exclude the use of glasses, and sintered alumina (Al_2O_3) is used instead. The exterior bulb is made of borosilicate glass.

Spectral lamps are used to produce line spectra and consist of a discharge chamber with a base glass and the material to be excited, as well as an exterior bulb made of soft glass. Quartz glass or special glass is used for the discharge chamber, depending on the type of discharge and the metal vapor filling used.

High transmission at the mercury line (254 nm) is required for UV-radiators which are used in sterilizers and other medical equipment. Quartz glass is therefore used. Radiation below 280 nm is undesirable in UV-radiation used for cosmetic-therapeutic purposes. Soda-alkali earth glasses are used in such instances. Their UV transmission can be controllably reduced by adding Fe_2O_3 to the composition. The shortwave UV-radiation can also be suppressed by a thin oxide coating applied to the bulb.

Soft glasses are used in *high voltage fluorescent tubes* containing inert gases such as neon. These are the familiar neon signs of advertising. The lamps contain small quantities of mercury to excite the gaseous discharge. Since UV-radiation and mercury attack and discolor normal glasses at the high operating voltages (up to 6kV), special glasses must be used for this purpose.

Soft glasses and borosilicate glasses are used in the production of *flash bulbs*. The glass used depends upon the temperature and pressure resulting from the flash.

Colored lamps can be made in different ways. Body colored glasses use CdS for yellow, Co and Cu for blue, Cr for green, and Se for red. Surface-tinted glasses can also be made. They are stained with silver or copper compositions to obtain yellow, red, and brown. (See Painting, Chap. 5.) The inner surfaces can also be coated with colored powders, or colored lacquer can be applied to the outer surfaces.

Infrared lamps are incandescent lamps having low filament temperatures and therefore emit a relatively high quantity of infrared radiation. They are usually made of borosilicate glasses into which molybdenum or tungsten wire is sealed.

ELECTRODE GLASSES

In the control of chemico-technical processes, such as the monitoring of drinking water, in medicine, in dairies, etc., it is important to know pH values. These values are measurements of the hydrogen ion concentrations in solutions, which in turn serve as a measure of acidity and alkalinity.

Today, pH values are almost exclusively measured by glass electrode chains. At the interface between a glass membrane (approximately 0.5 mm thick) and liquid electrolytes (ion-conductors), potential differences (*Galvanic voltages*) are created. The potential differences between the glass and the aqueous solution are dependent upon the pH value and, in certain glass compositions, the alkali-ion concentration of the solution (sodium ion concentrations, in particular). This effect is exploited to measure pH, pNa, etc. of a solution using glass electrodes.

Since glass in itself is not an electrode, the two-phase glass/solution system is expanded into a multiphase system having metallic end phase (electron conductors). Figure 6-14 is a schematic drawing of

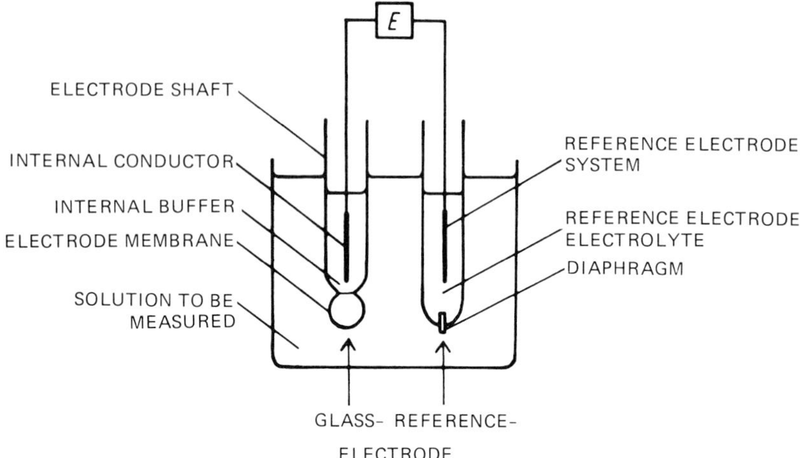

Fig. 6-14. Schematic diagram of a glass electrode measuring chain. The measured chain voltage E is composed of the voltages at the various phase boundaries (i.e., internal conductor/internal buffer, internal buffer/electrode membrane, etc.), which are all constant except the voltage of interest at the "electron membrane/solution to be measured" interface. Therefore, E is a function of the unknown pH value. The diaphragm prevents the mixing of the reference electrode electrolyte with the solution to be measured, but allows them to make electrical contact.

such a glass electrode measuring chain. Silver/silver chloride, calomel, and Thalamid® electrodes (Schott) are used for internal conductors and reference electrodes. The Thalamid electrodes have the advantage of good reproducibility and high operating temperatures (to 135° C).

OPTICAL AND OPHTHALMIC GLASS

Properties and Classification of Optical Glasses

The most important properties of glasses for optical purposes are their refraction of light and their dispersion of color, which are determined by their refractive index n_λ. The lower symbol, λ, expresses the dependency of the value n on the wavelength (λ) of the light. The refractive index of all glass increases from the red end of the spectrum towards the blue (i.e., it increases with reducing wavelength).

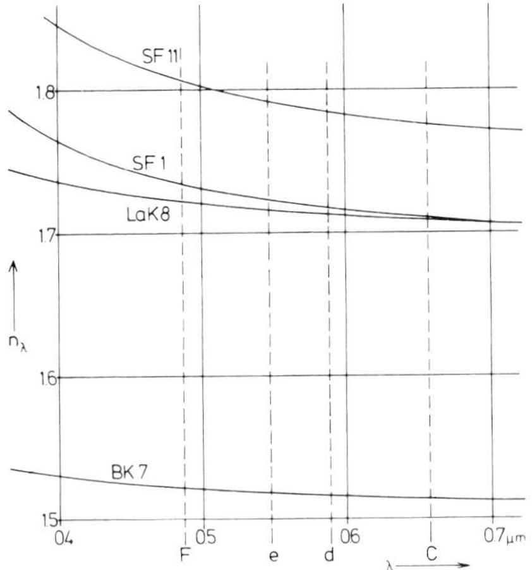

Fig. 6-15. Wavelength dependence of the refractive index (dispersion) for several optical glasses.

An increasing n denotes stronger bending of a ray of light striking the glass surface at an oblique angle. The curve of n_λ is identified by several standard spectral lines of chemical elements (Fig. 6-15). The most frequently used spectral lines are the yellow helium line d at 587.6 nm, the green mercury line e at 546.1 nm, and the blue and red hydrogen lines F and C at 486.1 and 656.3 nm, respectively. The difference $n_F - n_C$ is called the *average dispersion*. The ratio $(n_F - n_C)/(n_d - 1)$ is called the *relative dispersion*, and its reciprocal is called the *Abbe number* v_d. For optical identification, the optical glasses are entered on a chart, the abscissa of which is the Abbe number v_d, and the ordinate of which is the refractive index n_d. (See Fig. 6-16.) The position of a glass in this diagram is usually designated as its optical position.

For two glasses having normal dispersion behavior, it is usually valid that all dispersion values of the two glasses are compatible if their n_d and v_d values are compatible. For this reason, the whole refraction behavior of "normal glasses" are defined by these two values. A low Abbe number v_d signifies a glass with high dispersion, related to its refractive index. Such glasses are called *flint glasses*.

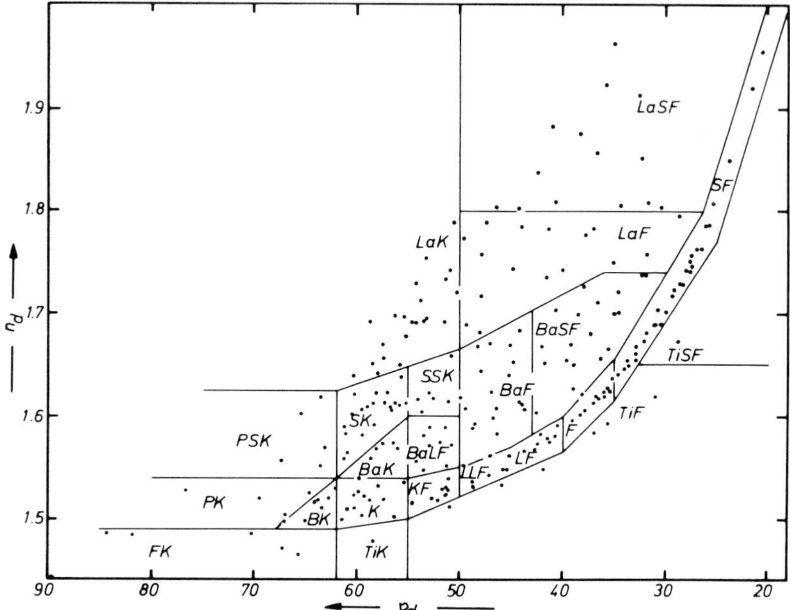

Fig. 6-16. Arrangement of optical glasses. Abbreviations: B = boron; Ba = barium; F = flint; K = crown; L = light; La = lanthanum; LL = double light; P = phosphate; S = heavy; SS = double heavy; Ti = deep. The points give the locations of the different normal optical glasses in this n_d/ν_d diagram.

Crown glasses, on the other hand, have relatively high Abbe numbers, related to their refractive indices. The Abbe numbers of the optical glasses lie between 20 and 90; the borderline between crown and flint glasses lies by definition at an Abbe number of 50. The historical development of the optical glasses followed in three successive steps.

Prior to 1880, optical systems consisted of simple crown glasses (soda-lime-silicate glasses) with relatively low dispersions, and simple flint glasses (lead-alkali-silicate glasses) with relatively high dispersions. Extensive glass trial melts using BaO, B_2O_3, and KHF_2, conducted by Otto Schott between 1880 and 1895, resulted in many new optical glasses. Barium oxide results in unusually low dispersion at high refractive index. Boron oxide yields a low refractive index and very low dispersion. The use of fluorine instead of oxygen also lowers the refractive index and dispersion.

In 1930, a new wave of new glass developments began. This work resulted in optical glasses containing rare earths, especially lanthanum. These glass types expanded the range of optical glasses available with high refractive index and high Abbe values. They were decisive in eliminating image defects in lens systems.

The optical glasses are traditionally identified by combinations of the terms *crown* and *flint* with the terms *heavy* and *light* which denote high and low refractive indices. Supplementary designations are given to subgroups according to the chemical components which are important in determining their optical positions (for example, barium heavy flint [BaSF] or phosphorous heavy crown [PSK], see Fig. 6-16).

Barium crown glasses contain a large proportion of boron oxide and barium oxide, while their SiO_2 content is relatively low. Small additions of substances, such as aluminum oxide stabilize the glass against devitrification and weathering. Small amounts of aluminum oxide inhibit fluoride crystallization in fluorine crown glasses. Fluorine, which volatalizes easily, can be added to the glass as potassium-bifluoride.

The borosilicate crown glasses correspond to the technical borosilicate glasses. The calcium oxide of normal soda-lime silicate glass is replaced by boron oxide.

The light and heavy flint glasses and barium crown glasses are characterized by low and high lead- and barium-content. Barium flint glasses contain barium oxide and lead monoxide; crown flint glasses contain calcium oxide and lead monoxide, resulting in average dispersions. The density of the glass decreases with increasing temperature, while the dispersion curve of the glass and its UV-absorption both shift. (See Fig. 6-15.) Both effects affect refractive index, so that it is possible to have glasses with either negative or positive temperature coefficients of n_λ. *Athermal glasses* are recent developments in which the optical light path ($n_\lambda \times$ glass distance) is nearly independent of the temperature. This is important in systems which have large focal lengths, such as telescopes.

Especially important for the manufacture of multilens high quality systems are optical glasses with which it is possible to construct not only achromatic but also apochromatic combinations. In an achromatic combination, the chromatic aberration is removed in only two colors (such as red and blue). However, a residual error (*secondary spectrum*) remains in the intermediate range. On the other hand, in an apochromatic lens, this residual error is almost completely corrected as well. This is brought about by using special glasses with different partial dispersions. Although these glasses have the same n_d and n_v values as the corresponding flint glasses, their increase in refractive index from the blue to the ultraviolet is "shorter" than in the corresponding flint glasses (hence, the name *short flint*).

Glass types with greatly differing partial dispersions are of particular interest in large diameter lens systems with long focal length. The resolution of fine details is so greatly increased in such systems that even the slightest chromatic defect is noticeable.

The quality demanded of optical glasses today far exceeds that of most other glasses. Among the most stringent requirements are freedom from striae, optical homogeneity (meaning consistency of refractive index within a melt), the smallest possible number of bubbles, minimal absorption in defined spectral areas, and low birefringence caused by internal stresses in the glass. Today, it is possible to produce highly homogeneous glass having a refractive index varying within $\Delta n_d \leq 1 \times 10^{-6}$.

For special applications, glasses are sometimes required to have refractive index behavior which varies in a prescribed manner. One such instance might be a rod having an index of refraction which decreases radially from the axis. Such *gradient rods* and *gradient lenses* are made of special glasses by ion exchange (Safety Glass, Chap. 4) or other techniques. Most of them possess a rotationally symmetrical index gradient. They are used in medical endoscopy, and miniature cameras.

Radiant Transmission; Color Filters

Transmission, absorption, and reflection are qualities which are equally important in optical glass, colored glass and filters, and colored ophthalmic glasses. Light, which strikes a sheet of glass is partially reflected, partially absorbed, and partially transmitted. (See Fig. 4-11.)

The degree of reflectance r is the ratio of the intensities of the light reflected from a surface and the incident ray of light. The total reflection R of a sheet of glass approaches $2r$ only if r and the absorption are very small.

The degree of transmittance d is the ratio of the intensity of light transmitted through the glass to that of the incident light. This results in the degree of absorption $A = 1 - R - D$.

The reflection losses of simple soda-lime-silicate glasses at perpendicular incidence are approximately 4% at each surface. However, they increase strongly when the angle of incidence increases above 40° to almost 100% at a glancing angle (90°). Absorption occurs when the quantum energy of the light is sufficient to excite the electrons bonded within the structure of the glass. In quartz glass and soda-lime glasses, the bonding of the oxygen atoms to the silicon or to the cations is so strong that electron excitement can only be caused by UV-radiation. So these glasses are completely transparent to visible light. Small concentrations of transition elements (see below) bring about not only strong absorption bands in the UV, but also broad bands in the visible and into the infrared spectral ranges. There are also several narrower absorption areas which can be attributed to OH groupings and water molecules within the glass. All silicate glasses are opaque to infrared waves above 5 μm as a result of the natural oscillation of the $Si - O$ groups which absorb these waves. Glasses for IR-optics cannot contain SiO_2 and they may contain no oxides whatsoever for wavelengths over 6 μm. In their place, compounds of such elements as sulfur, selenium, arsenic, antimony, and germanium are used. They must be manufactured in a completely oxygen-free environment. (See also Coloring Agents, Chap. 2.)

The colorings, which are the result of transition elements (such as Cu, Ti, V, Cr, Mn, Fe, Co, Ni, etc.), vary greatly. (See Raw Materials, Chap. 2.) The structure of the host glass also influences the absorption spectrum to a certain extent. Each change in the bonding relationships in the glass structure affects the electron bonding and therefore the color characteristics. (See Fig. 6-17.)

Fig. 6-17. Color filter glasses.

The addition of rare earths has a lesser influence on color. This is the basis for using such glasses (most of which contain neodymium) in generating laser radiation. (See Fig. 6-18.) Lasers (an abbreviation for "Light Amplification by Stimulated Emission of Radiation") are light sources in which the stimulation energy of a pulsed light source (xenon or krypton flash lamp) is transformed into monochromatic, coherent light of high intensity. They are used in such applications as "machining" of materials processing, measuring technology, optical data transmission, and eye surgery. The stimulated neodymium ions in the glass emit infrared radiation at the 1.06 μm wavelength.

Narrow band filters which are transparent to a narrow spectral range with half value widths <0.05 μm cannot readily be made with

Fig. 6-18. Laser glasses.

absorbing glasses nor with filters having a sharp transmission cutoff at longer wavelengths. However, such requirements can be fulfilled by interference filters which usually have many layers with different refractive indices. Some types also have several metallic coatings and are often used in combination with a filter glass.

Ophthalmic Glass (Spectacle Glass)

A number of special glasses are used for the correction of vision and to protect the eye against undesirable light radiation. They are usually melted with great homogeneity in electrically heated furnaces, most of which have a capacity of one ton.

Colorless ophthalmic crown glass (white spectacle crown) has an exactly defined refractive index of 1.5230. The curves of the lens surfaces and the refractive index of the glass determine the optical effect of ophthalmic glass. The refractive strength D is the reciprocal of the focal length (in meters) and is expressed in diopters (dpt.). Thus, an ophthalmic glass with a power of ± 2 dpt. has a focal length of 0.5 meters.

Between $20°$ and $300°$ C ophthalmic crown glasses have a linear coefficient of thermal expansion of 9.2×10^{-6}/K. The viscosity temperature dependence is comparable to that of soda-lime glass. Both factors (thermal expansion and viscosity behavior) are important for determining the fusibility of this ophthalmic crown glass with special flint glasses for multifocal spectacle lenses (bifocal and trifocal glass). (See Fig. 6-19.)

As the power of the prescription increases (above ± 4 dpt.), the edge or center thickness of crown glass spectacles becomes very thick, creating unpleasant weight for the wearer. For this reason, lighter, high-index heavy flint glasses (Schott) were developed. They result in thinner lenses having the same refractive strength.

For safety reasons, the mechanical strength of ophthalmic glasses has been improved so much by thermal and chemical strengthening processes that several special glasses have 8 times the strength of normal ophthalmic glass.

The colored absorption glasses are separated into lightly tinted comfort glasses and sunglasses. Both types are ophthalmic crown

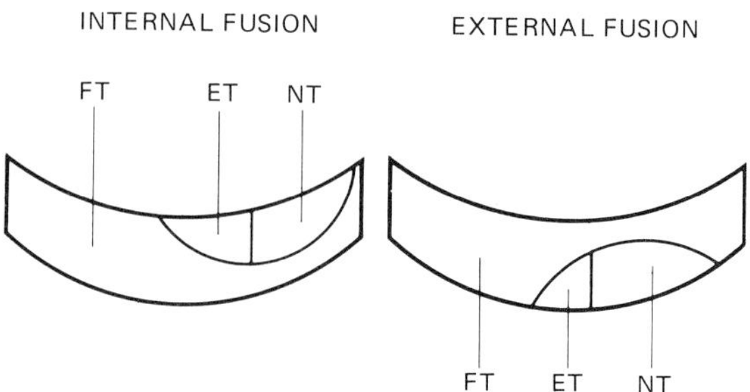

Fig. 6-19. Cross section of fused multifocal glasses. Internal fusion (left) and external fusion (right). In both processes, the reading segment NT, which has a high refractive index (higher than the major segment FT), is fused to the extension segment ET. Then, in a second step, both of these are fused to either the concave side or the convex side of the major segment FT.

glasses with 1.5230 refractive index, and most of them differ in coloring components only. The tinted absorption glasses, most of which are pink in color, have transmission values between 85% and 60% (based on 2 mm glass thickness). Most sunglasses are browns, greys, and greens with transmission values between 65% and 20% (based on 2 mm glass thickness). The coloring oxides used include those of iron, cobalt, nickel, copper, manganese (Raw Materials, Chap. 2), as well as combinations of iron oxide and selenium oxide. Fashionable tints, such as blue or yellow, complete the scale. Good sunglasses also have the lowest possible transmission of infrared and ultraviolet rays, which are harmful to the eye, though not perceived by it.

Safety glasses, which protect against thermal radiation, have strong concentrations of coloring oxides (Schott KG-glasses or athermal welding glasses made by Deutsche Spezialglas AG). These glasses have a defined vision-sufficient transmission within the visible spectrum, but are virtually opaque in the spectral regions harmful to the eye.

In photochromic ophthalmic lenses exposed to ultraviolet or short-wave infrared radiation, the transmission of visible light is automatically reduced. When the exposure is ended, they return to their initial state within a short period (Fig. 6-20). This property, called *phototropy*, is based on glassy or crystalline silver halide containing separations of submicroscopic size (diameter approximately 5-30 nm,

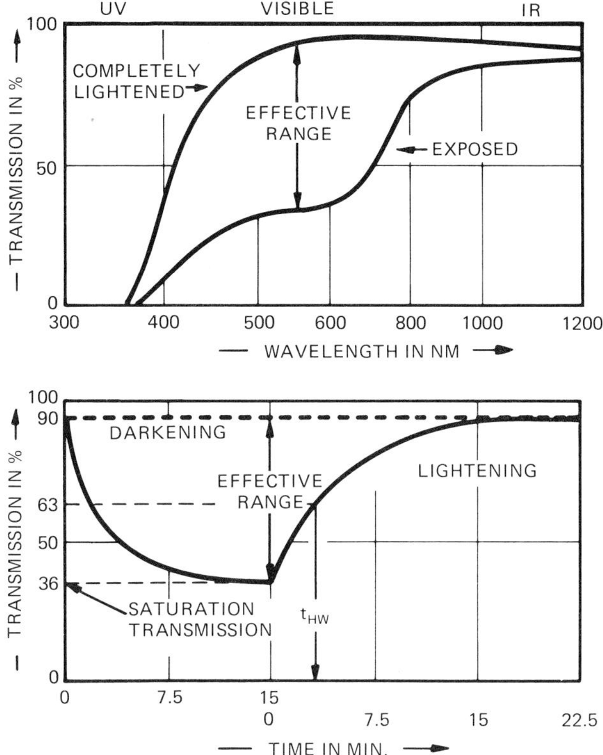

Fig. 6-20. Photochromic ophthalmic glass: (a) spectral transmission in the exposed (darkened) and lightened states; (b) darkening and lightening.

concentration approximately 1:2000). They are produced by adding silver salts and halides (metal compounds with fluorine, chlorine, or bromine) to the batch (usually a borosilicate-base glass). Closely controlled thermal treatment during and after the melting process causes the formation of the silver halide containing separations which bring about the photochromic process. (See Fig. 6-21.)

The ophthalmic sector is one of the few areas where glass encounters competition from plastics. For simple, fashionably tinted antiglare spectacles and industrial safety eyeware, materials such as plexiglass (polymethylmethacrylate) or polycarbonates are used.

Fig. 6-21. Photochromic ophthalmic glass with a gradient tint (left—unexposed; right—exposed).

They are not suitable for corrective eyewear. The major ophthalmic grade plastic is CR-39 (abbreviation for Columbia Resin 39 made by the U.S. firm, PPG; chemical designation: poly-diethylene-glycol-diallyl-bicarbonate). It is mainly limited for use in weak prescriptions since otherwise the low refractive index would cause the lenses to be undesirably thick in order to accomodate the necessary lens curvature. In addition to good optical quality, this material has adequate wiping and scratch resistance. In order to make this last property more comparable to glass, plastic safety spectacles (partly CR-39) are coated with harder thin layers. These coatings are usually precipitated from solutions of organic silicon compounds (methyl-polysiloxanes are often preferred). In some cases, vacuum deposited glass has also been used. (See Glasses with Altered Radiation, Chap. 4.)

Special Optical Glasses for Nuclear Technology and Radiation Research

Radiation shielding glasses and *radiation-resistant optical glasses* form a separate group which was specially developed for nuclear technology. (See Fig. 6-22.) In the shielding windows in "hot cells," the high absorption of radioactive radiation by lead is exploited. (See Fig. 6-23.) Since these lead containing glasses tend to discolor under α- and β-radiation, they are usually stabilized with cerium oxide. Schott's popular RS 520 G 5 radiation shielding glass, for example, contains 0.5% CeO_2. A sheet of this glass has about the same absorption as a sheet of lead with half the thickness.

Optical lead glasses are also used in radiation research to detect and determine the energy of high-speed subatomic particles (electrons,

Fig. 6-22. Radiation shielding glasses.

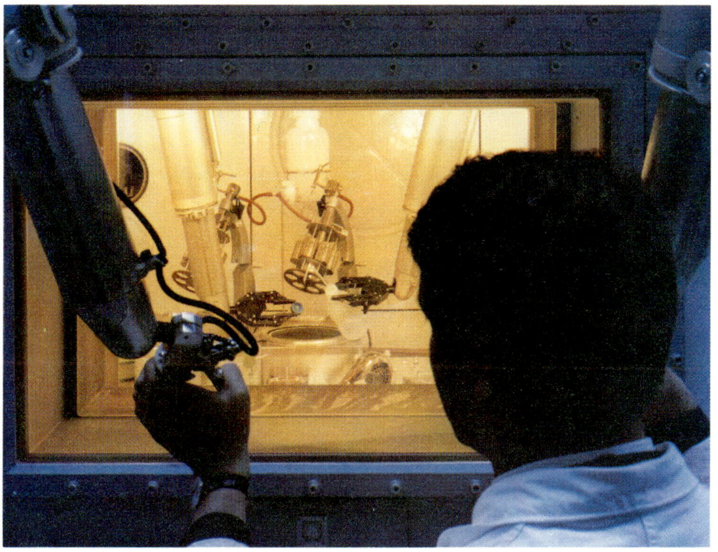

Fig. 6-23. "Hot cell" with a radiation shielding window.

positrons, cosmic rays, etc.). If such a particle travels through a medium (refractive index n) at a velocity of v which is faster than c/n, which is the velocity of light in the medium (c = the velocity of light in a vacuum, $= 3 \times 10^5$ km/sec), it forms a ball-shaped expanding electromagnetic head wave, the so-called *Cerenkov radiation*. (See Fig. 6-24.) This is visible as blue-violet light and can be registered by using photomultipliers and impulse counters. Special glasses with re-

Fig. 6-24. Optical glass blocks for Cerenkov counters.

fractive indices ≥ 1.7 and light transmission down to UV meet the requirements for measuring particle velocities $\geq 1.8 \times 10^5$ km/sec. Certain atomic nuclei are also capable of transforming photons from gamma radiation into electron-positron pairs which can also be verified by the Cerenkov effect. Similar methods of verification can be achieved by the use of *scintillation glasses*. By using the reverse effect of energy-rich radiation (particles or γ-quanta), certain atoms in glass are stimulated to localized light emission in the form of minute flashes (scintillations), which can be registered and evaluated by using highly sensitive multipliers.

Dosimeter glasses are used to measure dosages of γ-rays, especially small doses in personnel dosimetry. There are also special glasses for use in neutron dosimetry (threshold value detectors). Cobalt- or silver-containing phosphate glasses, which form color centers in radiation, are usually used. Cobalt-containing glasses color in radiation between 425 and 370 nm. Silver phosphate glasses produce silver particles when subjected to radiation. The intensity of fluorescence when stimulated with UV light is a measure of the disturbed centers and hence also of the radiation dose.

The Manufacture of Optical Glass

The older process for melting optical glasses chiefly utilized the ceramic pot (or crucible) which was heated alone or with up to 9 other pots in a gas-heated and mostly regenerative pot furnace. Glassmelts in large pots (up to 800 liters and above) were homogenized with stirring apparatus after the batch had been added and melted. After the desired homogeneity was obtained, the contents of the pot were cast into a rectangular steel mold to solidify prior to annealing. (See Fig. 6-25.)

Due to the low yield of such a process and the tendency of many optical glasses to devitrify, pot melting is hardly done anymore. Instead, larger quantities of optical glass are usually melted in continuous tank furnaces. Smaller quantities are melted in platinum or quartz crucibles (depending on the glass composition) in a method similar to the older pot process. Tank melting of optical glasses is essentially the same as tank melting of other kinds of glass. (See Chapter 3.)

Carefully controlled annealing is extremely important for optical glasses. In many cases, the annealing process can be used to make fine corrections to the refractive index.

As in pot melting, tank melting can also be used to produce blocks with the optical quality easily tested. In addition, continuously produced preshaped glass extrusions can also be made with rectangular, circular, or triangular cross sections. (See Fig. 6-26.) These can be processed into optical glass components by the optical processing industry. Rotary table presses in conjunction with suitable glass-feed mechanisms (e.g., automatic shears) enable blank prisms, lenses, and

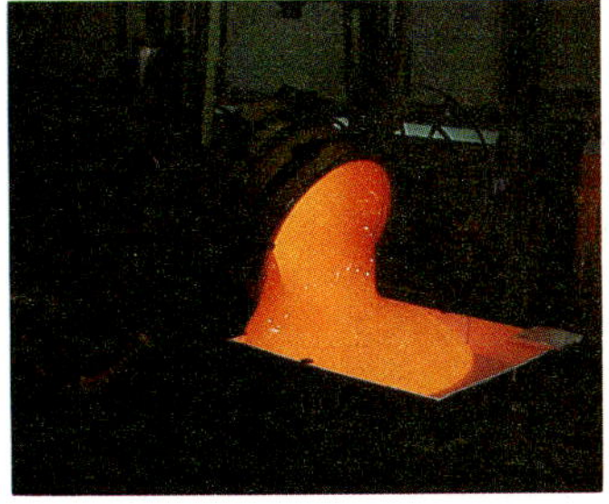

Fig. 6-25. Pouring pot melt.

Fig. 6-26. Raw optical glasses.

spectacle lenses to be made from molten glass flowing directly out of the melting tank. (See Fig. 6-27.)

Glass Spheres

Uses and Properties.   Clear and colored sphere-shaped glasses with different refractive indices in the form of small spheres or beads are not only used for fashion jewelry. There are also many technical uses for them.

Because of their optical qualities, microbeads (also called *Ballotini*) having diameters of ≤ 0.2 mm are used in reflective signs. They are also used in projection screens. Glass spheres used for this purpose have diameters between 0.10 and 0.16 mm and a refractive index of about 1.70. They are embedded in colored or fluorescent plastics or lacquer in order to increase their reflective qualities. Reflective foils having embedded glass spheres are widely used in traffic signs and road markings. The spheres act as optical lenses which concentrate incident light coming from a distance into a point behind but near to the back of the spheres. When many spheres having approximately the same diameter are placed next to one another, all illumination is focused onto a single plane behind the spheres where a highly reflective metallic film is located. The incident light is thereby reflected back through itself. (See Figs. 6-28 and 6-29.)

The distance of the focal point from the rear side of the sphere (f) is dependent upon the effective refractive index $N = n/n_0$ (n, n_0 are refractive indices of the sphere and the surrounding medium), and it decreases with increasing n.

Fig. 6-27. Automated pressing of lenses.

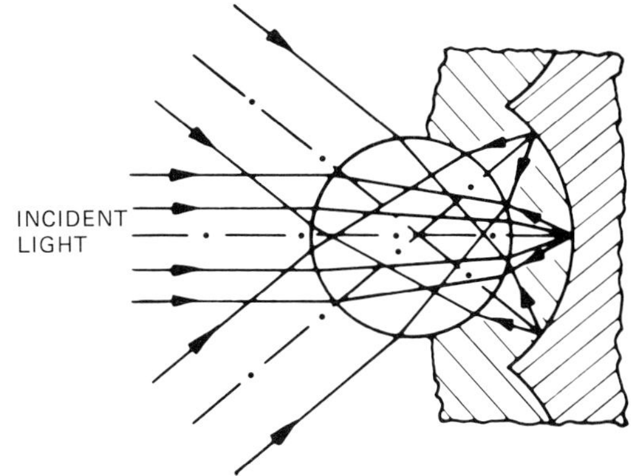

Fig. 6-28. Light path in a retroreflective system with glass spheres.

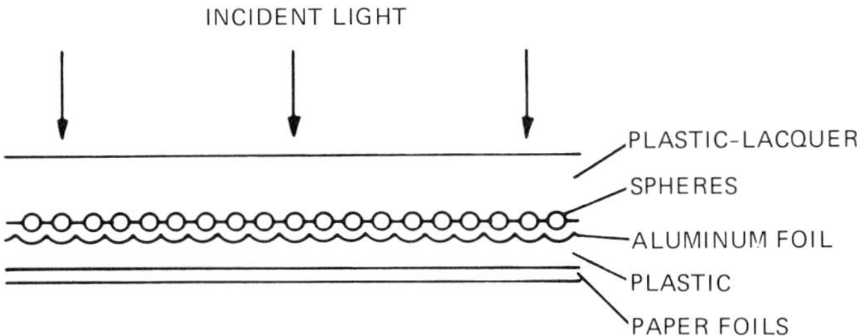

Fig. 6-29. Schematic drawing of a retroreflective foil.

For this reason, highly refractive glasses with n up to 2.3 are used in retroreflective foils. Such glasses are made from either lead-alkali-silicates or more chemically stable and leadfree barium-titanium-zirconium-borosilicate compounds.

Manufacturing Process. The industrial manufacturing methods used for glass spheres depend on the glass composition and the desired sphere size. For microspheres (diameter $\leq$ 0.1 mm) made of low melting temperature calcium-alkali-borosilicate or lead-alkali-silicate glasses (refractive index 1.5-1.65), the following manufacturing methods are used:

- The gravity shaft process is the simplest for low melting glass types. Presifted, fractionated glass powder of approximately the desired sphere size trickles through a heated shaft, where the individual particles melt and assume a spherical shape. The spheres are collected in a container below the shaft.
- In another process, presifted fractionated glass powder of desired particle size is fed into a furnace which has an upward directed flame. Under the influence of the temperature and the surface tension, the pieces assume spherical dimensions in the flames. The hot air stream forces them upwards into a cooler zone. The spheres solidify quickly in the cold zone and are collected in a suitable container.
- The high melting temperature barium-titanium glasses and the high lead content glasses with high refractive indices crystallize easily. Therefore, cooling to solidification must be very fast (quenching at $\sim 1000°$ C/s). This can be done by spraying a thin stream of glass under high pressure into a collecting chamber.

GLASS FIBER

Glass fiber is a collective term for glass processed into fibers which have diameters between 0.1 mm and a few thousandths of a mm. The development of techniques for drawing fibers from a glass mass has opened the way for additional, wide-ranging future applications.

For a long time, two main groups have been recognized—*insulating fiberglass* and *textile fiberglass*. Both have different applications and characteristics and are made of different types of glass.

A recent development is *lightguide fibers*, or optical fibers, which are quickly finding a variety of potential applications in science and technology.

Insulating Glass Fibers

These are normally made of soda-lime glass by the centrifugal process using a rotating disk. Glass droplets are drawn into endless fibers by centrifugal force (rotation). The fibers cool in air and are hardened. The same effect is achieved by airjets, whereby the airstreams cause the formation of fibers from molten glass. The individual fibers immediately matt together. This is sometimes called *glass wadding* and it can be used as it is or can be further processed.

The actual weight of loosely piled, insulating glass fibers is between 30 and 200 kilograms per cubic meter. Therefore, they are particularly suited for building insulation, since they add no significant load to the structure. Fiberglass insulation can be easily combined with other construction materials, such as mortar and plaster. The durability of the glass ensures a long life for fiberglass insulating products.

Products of Insulating Glass Fibers. *Glass wool*, another name for insulating fiberglass, is used in the form of loose glass wadding, or as strips, matting, felts, panels, dishes, and pressed parts. The fiberglass is either impregnated with a resin or layed between sheets of paper. Dishes and other molded parts are formed from the resin-impregnated glass wool. Soft matting or stiff panels are pressed.

The Uses of Fiberglass Insulation.   In construction, especially residential construction, it is mainly used for thermal and acoustic insulation. The fiberglass matting or panels are applied in exterior walls, and fiberglass wool is used as a layer in partitions or beneath floors thus resulting in considerable savings in heating costs. Pipes can similarly be shielded from the environment. The soundproofing ability of fiberglass insulation is due to the many small hollow spaces which impede the spread of sound waves. Fiberglass insulation is an indispensable product for modern environmental protection.

Fiberglass Textiles

These are also made of fine fibers produced from molten glass. The fibers are uniform, usually have a circular cross section, and can be further processed into threads. The diameter of the fibers is less than 18 μm. They are subdivided according to length and manufacturing method into glass filaments and glass staple fibers.

Glass Silk.   This consists of individual glass fibers (glass filaments) which are between 5 and 18 μm in diameter and of any length. They are spun into glass threads.

To manufacture glass fiber-filament yarn, a special batch is melted in a tank. The molten glass flows in channels, and filaments are drawn through a platinum jet at high speed and rolled on a drum into the finest of threads (filaments).

Glass Spun Threads.   Before the glass fibers can be wound into a "cake," the filaments are wound in parallel spun threads which can consist of 50 filaments or multiples thereof. Next, the spun threads are spooled off the cake into a so-called roving skein as a commercial product. (See Fig. 6-30.) The strength of freshly drawn glass spun fibers is as much as 50% of the theoretical strength of defect-free glass. (See General Characteristics, Chap. 2.) If several spun threads are wound together, glass silk yarn results. And glass silk twine is made by winding several glass silk yarns together.

Fig. 6-30. Spool rack for glass staple fiber yarn.

Glass Staple Fibers.   These are glass textile fibers of a certain length and diameter. In the airjet manufacturing process, glass flows through orifices in the bottom of a platinum melting tank. Compressed air blows the glass into filaments between 3 and 60 mm in length. They are sucked through a perforated drum from which they are drawn into glass staple fibers.

Due to the manufacturing process and partially due to diagonally crossed fibers, these fibers are only half as strong as glass spun fibers. But because they are cheaper to make, they are preferred to glass silk yarn in certain cases.

Glass Types Used in Making Fiberglass Textile.   The most widely used type is the so-called *E-Glass*. It is essentially alkali-free (sodium and potassium content less than 0.8%) and is characterized by water resistance and high softening point. E-glass is an alumino-borosilicate glass by composition. If acid resistance is required, then *S-glass* is used. This is an alkali-lime glass with a large percentage of added boron, primarily utilized in glass staple fiber weaves for corrosion protection, sealing, and surface protection. *D-glass* is used if dielectric qualities are important; *R-glass*, for more stringent mechanical requirements. *M-glass* has a high elasticity modulus.

Processing Aids.   Untreated glass spun threads and glass staple fibers may not be further processed because of their surface sensitivity. Hence, substances are applied to their surfaces before further processing. Dressings and lubricants are organic substances with which the glass fibers are coated immediately after they are drawn. This serves to minimize the scouring effect of glass-to-glass contact and limits the danger of mechanical damage which could occur during spinning. In glass spun fibers, the additive is called a *dressing*, in glass staple fibers, a *lubricant*.

Adhesive Agents.   These are needed to improve the adhesion of synthetic resins and textile glass and glass fiber reinforced plastics.

Adhesive agents are mixed with either the dressings or lubricants. They promote the internal adhesion of both materials, and they reduce shrinkage of the plastics on setting.

Fiberglass Reinforced Plastics.   High tensile strength of up to 10^3 N/mm² and low expansion are the properties which render fiberglass suitable for improving the mechanical qualities of plastics. The strength of fiberglass reinforced plastics increases as the glass content increases. The fibers are added to the plastic in the form of spun threads, bands, strings, and cloth. Glass fibers imbedded in the plastic assume the mechanical load placed on the material before it is transferred to the plastic itself. Impact resistance and absorption of vibration enable fiberglass reinforced plastics to be used in protective helmets, chassis parts, telephone booths, loud speaker housings, etc. (See Fig. 6-31.) Unlimited colors are available, and because of excellent weathering characteristics surfaces do not need to be lacquered. Depending on the difference in refractive index between the fibers and synthetic resin, transparent or translucent skylights, lighting fixtures, or level identification storage tanks can be made. The material is also resistant to many chemicals, so that it can be used to make acid containers, galvanizing troughs, piping, fodder silos, and even beverage and food containers.

Glass Fiber Optics

This term covers several high-quality products which are made of fiber light guides.

Fig. 6-31. Fiberglass reinforced plastic canoe.

Fiber Light Guides.   Optical fibers, or fiber light guides, are optical systems for the guidance of light and the transmission of images over paths which can turn in any direction. (See Fig. 6-32.) A fiber light guide is a thin, flexible filament with a diameter of a few hundredths of a millimeter. The interior or *core* consists of highly refracting optical glass surrounded by glass of lower refractive index, called the *sheath*. If a light beam strikes the end of a fiber within a defined incident angle determined by the refractive index difference between core and sheath, it is guided by total reflection at the interface between core and sheath to the opposite end of the fiber. The sheath optically insulates the cores from one another, so that light from one fiber may not be transmitted to an adjoining fiber. Fiber light guides which operate according to this principle are also called *stepped index fibers*.

The Manufacture of Light Guide Fibers.   In the rod-tube process, a rod made of high refractive index glass is placed inside a tube made of low refractive index glass. When heated in a furnace, both glasses soften and are simultaneously drawn and wound on drums or spools. In the two-crucible process, both glasses are melted separately before they are drawn through a concentric dual orifice into a fiber and then wound. (See Fig. 6-33.)

Light Guides.   These are bundles of light guide fibers which are either completely attached to one another (rigid light guides), or only attached at the ends (flexible light guides). (See Fig. 6-34.) At the ends, they are ground and polished perpendicularly to the fiber axis.

Fig. 6-32. Optical fiber light guides.

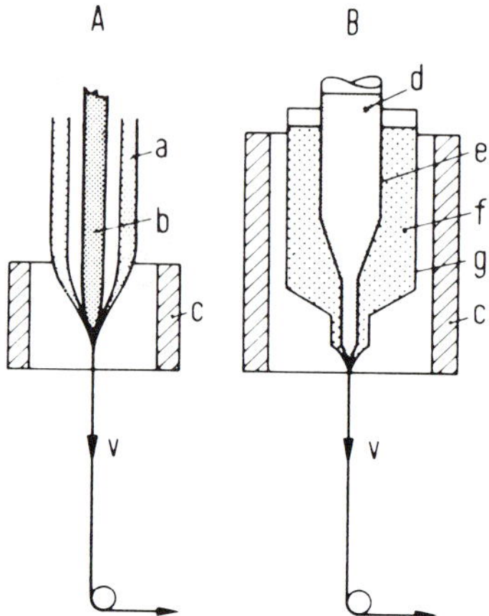

Fig. 6-33. Manufacture of light guides and fibers made of a core glass and a sheathing glass: (A) the rod-tube process, (B) the two-crucible pot process, (a) glass tube, (b) glass rod, (c) ring furnace, (d) core melt, (e) inner crucible, (f) sheathing melt, (g) outer crucible.

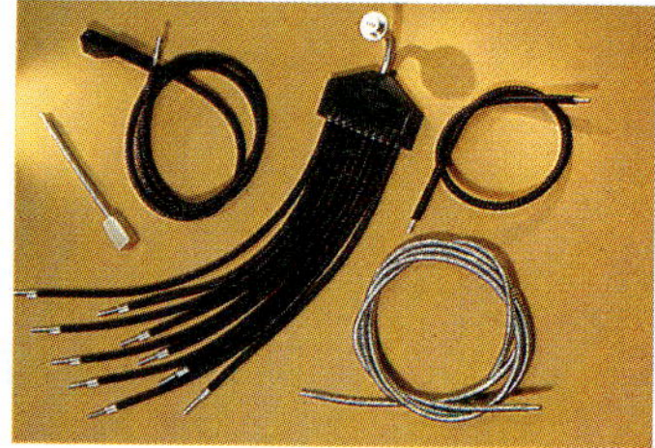

Fig. 6-34. Light guide cables with glass fibers.

In order to protect them from outside influences, the fiber bundles are enclosed in flexible cladding or rigid housings.

Image Guides.   These consist of bundles of light guide fibers with ends arranged in the same configuration. In this manner, images are picked up as individual points of a grid and transmitted. Each fiber transmits only one point of the image. The smaller the diameter of the fibers, the better the resolution (the separation of neighboring image points). (See Fig. 6-35.)

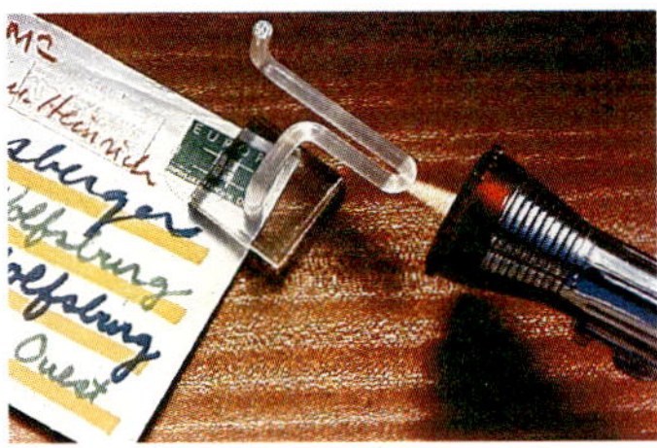

Fig. 6-35. Rigid image guides.

Light Guide Cables.   By covering fiber bundles with plastics, light guide cables can be made continuously and wound on drums. They resemble electrical cable. Shorter lengths are sometimes covered with metallic cladding instead of plastic. Cables are cut to the desired length with the ends packed as tightly as possible, set into metal housings, and then polished. They are relatively inexpensive if used in great quantity.

Multiple branch light guide cables can distribute light from a single source to many points, or they can concentrate light from many sources at one point. The opposite ends of the fiber bundles can also have different diameters, thereby altering the density of the light.

The Uses of Glass Fiber Optics.   Light guides are used where normal light sources cannot be used. They are used for such medical purposes as endoscopy, whereby cold light can be used to illuminate internal human organs. *Cold light* is obtained by filtering out the thermal components of light before they enter the light guide. Flexible fiber bundles are also used for illumination microscopy and for close work illumination such as in the restoration of ancient works of art. The reading heads in computers contain fiber optic light guides, as do control panels, changeable traffic message signs, and oil furnaces (the light from the flame is transmitted to a shielded photocell). (See Fig. 6-36.) In changeable traffic message signs, which are illuminated by a central light source, the many branches of a multibranch light guide are arranged to form words or numbers on a panel. Fused fiber optics plates of approximately 10 μm diameter and just a few millimeters in

Fig. 6-36. Changeable traffic message sign using light guide fibers.

just a few millimeters in length even image distortion in electrooptical picture tubes. Image converters and residual light amplifier tubes having fiber optic plates have greatly increased night vision technology. UV-transmitting light guides, with cores made of quartz glass and sheaths of low refraction plastic are used in dentistry for treating tooth-decay.

Fiber Light Guides Used in Communications Technology. The most important future use of fiber light guides is undoubtedly in communications technology. They enable the rapidly increasing demand for telephone and television transmission capacity to be fulfilled, while at the same time replacing even scarcer copper. The electrical signal impulses of the transmitting station are converted by semiconductor devices into light (or infrared) impulses and are transmitted over great distances without having to be amplified along the way. At the receiving end, photodiodes and amplifiers again transform the light impulses into electrical impulses which control the telephones or televisions. A prerequisite for viability was drastic reduction in optical losses in the glass. While the light intensity falls to 50% after traversing one meter in normal quality optical glasses, special processes have increased this distance to one kilometer for optical fibers. This corresponds to a loss constant of about 3 dB/km. In order to achieve a transmission capability of 1 gigabit $\times$ km/sec*—which is

---

*bit = one element of information (0 or 1); 1 gigabit = 10^9 bits.

sufficient for handling 20,000 telephone conversations or 20 television programs—gradient index fibers are needed. (See Properties of Optical Glasses, above.) These have a refractive index falling from the core center to the edge of the fiber according to a predetermined mathematical function, thereby producing a uniform optical path length for all modes, i.e., for all light rays irrespective of their angle to the fiber axis. Tests on pilot installations of such a system have already been successfully concluded. (See Fig. 6-37.) In the case of a step index fiber, light rays tracing zig-zag paths through the fibers must travel a longer distance than rays travelling straight down the center. Because of this, not all light arrives at the same time and the fiber is said to have a high signal dispersion. In gradient index fiber, light rays travelling away from the central axis pass through regions having progressively lower refractive indices. Because the velocity of light in a medium is inversely proportional to the refractive index,

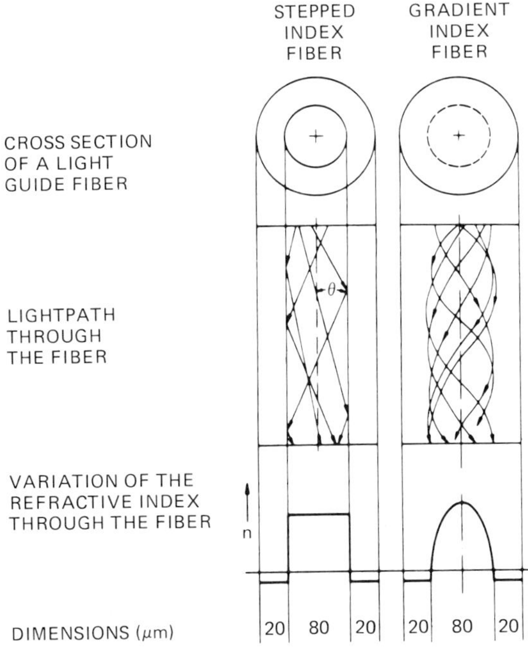

Fig. 6-37. Comparison of different refractive index profiles in stepped index fibers and gradient fibers.

skewed rays travel faster than light travelling down the center of the fiber. When the right refractive index gradient is used, the net result is that virtually all light arrives at the same time, and signal dispersion is greatly reduced. Gradient index fibers are presently the material of choice of telecommunication fibers.

The refractive index gradient is produced by heating a quartz glass tube to a very high temperature and passing through it a carefully dosed vapor mixture stream of compounds of glass-forming elements (Si, B, Ge, P) together with oxygen. The resulting glassy oxides precipitate on the hot inner wall of the tube in a preprogrammed relationship. Then the tube is heated until it collapses. The fiber is drawn from the resulting "preform" and maintains the same refractive index gradient, but on a smaller scale. (See Fig. 6-38.)

So called single-mode or wave guide fibers are being discussed as alternatives to gradient fibers. They consist of a sheathed homogenous fiber core of a few micrometers in diameter and can also transmit at high bit rates, but are much more difficult to install and adjust.

High transmission capacity, small space requirements, and the total insensitivity to external electromagnetic influences make optical communications fibers an ideal transmission medium for modern telecommunications.

GLASS CERAMICS

As already mentioned in Chapter 2, all glasses are in a supercooled state when cooled below the melting point of crystals which have the

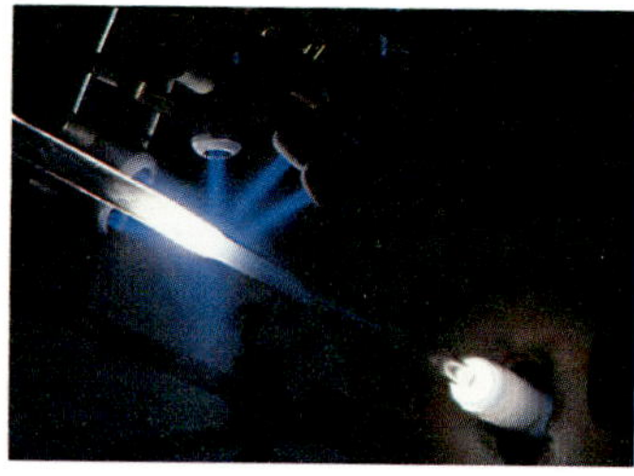

Fig. 6-38. Manufacturing the preform for gradient fibers.

same chemical composition. Crystallization (devitrification) does not occur because crystalline growth, which is controlled by the diffusion of the components, is too slow as a result of the rapidly increasing viscosity of glass due to falling temperature or because the number of nuclei, from which crystallites (smallest particles identifiable as crystals) can be formed, is too small. In glass ceramics, on the other hand, crystalline growth is deliberately stimulated in suitable glass systems in order to obtain materials with special properties.

The starting point is a glassmelt from which the desired items are formed by pressing, blowing, rolling, or casting. During a subsequent heat treatment which follows a prescribed temperature-time curve (Fig. 6-39), submicroscopic crystallites begin to form. The addition of high melting point materials (usually TiO_2 and ZrO_2) to the melt is necessary. When they precipitate, they act as nucleators and set the crystallization in motion. In doing this, it is important that the temperature zone of nucleation (T_{Kb}) lie below the temperature zone of maximum crystalline growth rate (T_{Kr}). (See Fig. 6-40.) The glass cannot then crystallize when the melt is cooled as long as there still are no nuclei. Only when a sufficient number of nuclei have formed at T_{Kb} can the minute crystallites be formed in large quantities (up to $10^{17}/cm^3$) by reheating to T_{Kr}. The percentage of crystals by volume can be between 50% and 90%, depending on the qualities desired.

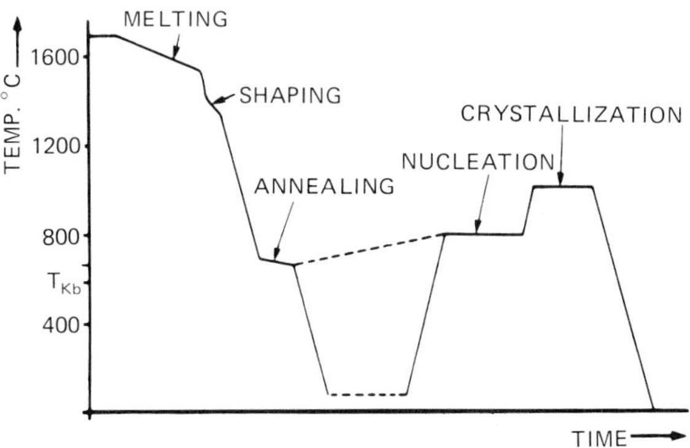

Fig. 6-39. Principle of temperature treatment of glass ceramics.

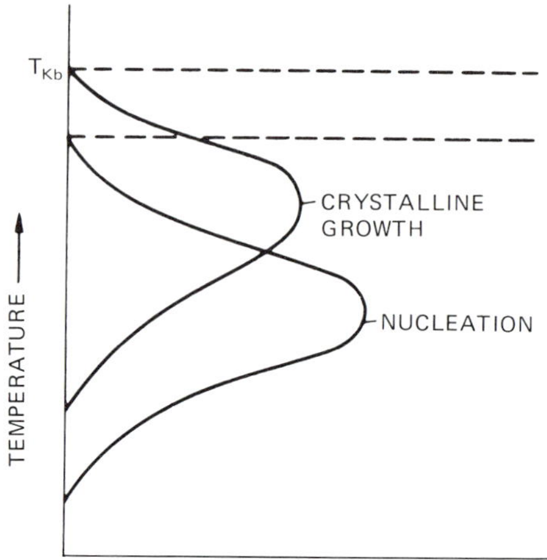

Fig. 6-40. Temperature dependence of nucleation and crystalline growth (schematic) rates.

Glass ceramics have technical importance in that their characteristics are not determined by their glass content only, but also by the types of crystals they contain. Some systems, which have crystalline phases with very low or even negative thermal expansion (such as lithium-alumino-silicates), are very important. They form materials with near-zero expansion over a wide temperature range, maintain their shape up to 800° C and are totally insensitive to thermal shock. They are used for stove panels, cookware, mirror substrates for telescopes, length standards, etc. (See Fig. 6-41.)

In another group of glass ceramics, the components of which are Si, Al, Mg, K, F, and O, micalike crystals appear during the ceramic-

Fig. 6-41. Telescope mirror substrate blank made of *Zerodur* glass ceramic.

making process. They exhibit a low degree of brittleness and allow the glass ceramic to be machined on lathes (*Macor*, Corning).

Between glasses and ceramics there are several photosensitive glass types in which crystallization can be prepared by UV-radiation and set in motion by subsequent heating. The presence of several percent of alkalifluoride, zinc-, and aluminum-oxide, as well as slight amounts of silver compounds ($\leq 0.2\%$) and cerium oxide ($\leq 0.05\%$) in the silicate glass matrix is necessary to achieve this effect. UV-radiation causes the electrons from the cerium ions to migrate; and they are trapped by the silver ions when heated; thus forming metal atoms which immediately transform into metal colloid particles. They act as nuclei for the devitrification process. When further heated, the system transforms into a yellow or brown glass ceramic. In this way, UV light can be used to transfer and fix pictures in the glass.

The presence of additional halides (bromine and chlorine) allows different colors to be fixed in the glass. These new *polychromatic glasses* (Corning) will find technical and decorative use in the future.

It is also important that the small crystals can be etched out at least ten times as fast with hydrofluoric acid as can the surrounding glass matrix. These qualities enable the production of high-precision etched shapes such as perforated plates for displays, printing dies, and the like. (See Fig. 6-42.)

FOAM GLASS

In the description of the batch melting process (Melting Process, Chap. 3), the high gas content of the primary melt was mentioned. It

Fig. 6-42. Chemically etched components made of light-sensitive glass ceramics.

was further pointed out that bubbly glass would result if insufficient refining took place. If the opposite approach is used, and additional gases or gas-bearing materials are added to the melt, the quantity of bubbles can be increased to make foam glass.

In another process, powdered soda-lime glass is mixed with pulverized sulfates, and charcoal is filled into molds and sintered. During melting, the mixture reacts to form SO_2 and CO gas. The gas bubbles, which are initially spherical, assume a honeycomb structure containing polyhedral cells under the influence of increasing gas pressure. The resultant foam glass has high compressive strength and dimensional stability with a density of only 0.13 to 0.3 g/cm³. The cellular structure means that the thermal conductivity of the product is about ten times less than that of a compact glass. These properties make foam glass particularly suitable for use in thermal and acoustic insulating materials for building construction. Foam glass can also be coated with concrete and installed in floors. Due to its low density and water impermeability, it is also used in floatation devices.

A GLANCE INTO THE FUTURE

The knowledge gained and developments made in glass technology within the past few years suggest that the possibilities have not yet been exhausted for finding glasses with special properties for new applications or for finding new technologies to make them. Indications of this were described in "Electron Conductive Glasses" in Chapter 6. Several years ago, it was discovered that certain oxide-free metallic compounds can be converted into a glassy state if they are quenched extremely fast (10^5 K/sec). Such cooling rates can only be achieved in thin ribbons or wires. In the United States, several alloys are already being processed into narrow glassy ribbons by using continuous casting processes. Precious metals and transitional metals, such as Pd, Fe, Ni, and Cr combined with one or more metalloids (Si, P, B, C) are especially suitable for this. Such typical metallic properties as flexibility, high strength, electrical conductivity, and even ferromagnetism remain despite the amorphous (noncrystalline) structure of the products.

Investigations of new glass manufacturing techniques are also being conducted with the object of making glass without high-temperature melting. One promising method uses silica gel solutions with additions of soluble compounds of metals or hydrolyzable alkoxides of several glass forming elements. They are slowly heated and with the efflux of water are brought to reaction and polycondensation. This principle has long been in use for producing special thin film coatings on glass by dipping, but no viable process yet exists for making larger glass components. Success in this direction could mean a great step forward for glass technology.

7. Glass as an Economic Factor

In all industrialized countries, the glass industry is among the smaller branches of production. At most, its value is one percent of total production.

In 1980, world glass production was valued at over DM 75 billion with an annual output of about 68 million tons. The Federal Republic of Germany accounted for about 7% of the total. The largest glass producing nations are the United States (over 25%), the Soviet Union (9%), and Japan (8%). Of the total value of world glass production, flat glass accounts for about 25%, hollowware 60%, special glass 10%, and fiberglass 5%.

The 340 or so glass-producing, finishing, and processing firms in West Germany employ approximately 77,000 people.

Exports average 24% of total production. Some export-intensive branches export a greater proportion of their output. These branches include tableware and special glass, with approximately 40% of their total output exported to more than 100 countries. The product range of the German glass industry is matched in its comprehensiveness only by those of the United States and Japan. It has a key position in the West German economy. The glass industry is very closely related to many large industries. Glass is a component of many of their own products. At the top of the list of consumers of glass is the construction industry. It is followed closely by the food and beverage industry. Other consumer groups are households, catering, the automobile industry, the electrical industry, plastics and textiles, chemical and pharmaceutical industry, medicine and research, furniture industry, and the optical industry.

The European countries are the classical glass producers. This is historically based. When the Industrial Revolution began in the eighteenth century, it also embraced glass production which had hitherto been exclusively manual. The combined glass industries of the Common Market approach the production potential of that in the United States. There is also extensive glass production in the East European countries. The different economic structure of these countries with their production plans and import/export monopolies (state trading companies) often presents considerable problems to the competing glass companies of the market oriented economies.

Developing nations are also increasing glass production capacities, usually obtaining financing and technical support from the industrialized nations. In so doing, they create jobs and supply domestic needs for glass products which are relatively easy to produce. The worldwide availability of raw materials for glass encourages this tendency.

Appendix

GLASS MUSEUMS

Most art history museums have glass collections. In addition, the following museums give an overview of glass history, production processes of the past and present, and the most important glass products:

Glasmuseum Frauenau
8371 Frauenau bei Zwiesel/Bayerischer Wald

Glasmuseum Wertheim
Muehlenstrasse 24
6980 Wertheim/Main

Kunstsammlungen der Veste Coburg
8630 Coburg

Museum of Glass
Corning, New York 14830

The Toledo Museum of Art
Toledo, Ohio 43697

EXPLANATION OF PHYSICAL SYMBOLS AND UNITS

Physical and technical data are always expressed as a product of a numerical value and a unit. The *fundamental units* (meters, seconds, temperature, etc.) are identified by the symbols m, s, K (Kelvin), and so on. Data, the numerical value of which can extend to many powers of ten, are prefixed by symbols distinguished from one another by a factor of one thousand (10^3). The most common used are:

k (kilo) = 10^3	M (mega) = 10^6	G (giga) = 10^9
m (mille) = 10^{-3}	μ (micro) = 10^{-6}	n (nano) = 10^{-9}

The often-mentioned *unit of length* μm therefore means 1/1000 millimeters. As *time units*, seconds (s) and hours (h) are frequently used. Frequencies are expressed in Hz (Hertz) = c/s, kHz (10^3 c/s), MHz (10^6 c/s), etc.

The *absolute temperature* T (in K) = t (in °C) + 273.15. Therefore, temperature differences in K have the same value as differences stated in °C. The *coefficient of thermal expansion* α of a solid body states the relative linear expansion per K (1° C) temperature increase, which in most glasses is in the order of 10^{-6}/K.

The technical *unit of power* is the watt (W), along with kW, MW, etc. Therefore, the *unit for energy* is the watt second (Ws) (also called the Joule), and the kilowatt hour (kWh). Since mechanical energy (work) is defined as the product of force times distance, the *unit of force* N (Newton) can also be expressed by Ws/m.

A force exerted on a solid body produces according to its direction either *tensile or compressive stress* σ, which is measured in N/m^2 = Pa (Pascal), and which also serves as a *measure of strength* in determining the maximum allowable load. These units (preferably N/mm^2) are also used to express the *modulus of elasticity* E which is used in determining the relative elastic expansion ϵ according to Hooke's Law ($\epsilon = \sigma/E$).

The relationship between transverse force and speed of a plate moving in a viscous liquid (melt) results in the *viscosity coefficient* η which is expressed in Pascal seconds (Pas); the old unit, Poise, is equal to 0.1 Pas (1 deciPas).

ATTENUATION OF RADIATION

As radiation (light, X rays, electrons, etc.) traverses a distance s through matter, it decreases in intensity (I) due to absorption or dispersion, to a value $I_0 e^{-\beta \cdot s}$. The material constant β has a dimension of m^{-1}, since $\beta \cdot s$ ($= -\log_e I/I_0$) is a dimensionless number. It is customary, however, to identify such logarithmic ratios with the unit *decibel* (dB) (0.1 Bel) when they denote attenuation. Usually, dB-values are referenced to s, whereby for example 1 dB/km corresponds to the value $\beta = 2.303 \times 10^{-4} m^{-1}$. (See Fiber Light Guides, Chap. 6.)

Similarly, dB is used in acoustics to identify sound strength (audio volume). Since the human ear perceives sound strength (P) more or less proportionally to its logarithm, the sound strength level has been defined as $L = 10 \log_{10} P/P_\circ$, where $P_\circ$ is the smallest perceivable audio volume at 1000 Hz ($\approx 10^{-12} \text{W/m}^2$). Thus for example, $P/P_\circ = 10^5$ results in an audio volume of 50 dB. This number approximates also to the number of *phon* at the measurement location. (See Glasses With Altered Radiation, Chap. 4.)

TECHNICAL LITERATURE ON GLASS

H. Jebsen-Marwedel: Glas in Kultur und Technik; Verlag Auman KG, Selb, 1976.

L.B. Klindt und W. Klein: Glas als Baustoff; Eigenschaften, Anwendungen, Bemessung; Verlagsgesellschaft Rudolf Mueller, Cologne 1977.

Sigurd Lohmeyer: Werkstoff Glas; Sachgerechte Auswahl, optimaler Einsatz, Gestaltung und Pflege; Expert-Verlag GmbH, Grafenau 1979.

R. Persson: Flat Glass Technology; Plenum, New York, 1969.

L.D. Pye etal, eds.: Introduction to Glass Science; Plenum, New York, 1973.

H. Scholze: Glas-Natur; Struktur und Eigenschaften, 2nd Edition 1977, Springer Verlag, Berlin.

W. Vogel: Glaschemie; VEB Deutscher Verlag fuer Grundstoffindustrie, Leipzig, 1st Edition. 1979.

Index